SpringerBriefs in Applied Sciences and Technology

PoliMI SpringerBriefs

Springer, in cooperation with Politecnico di Milano, publishes the PoliMI SpringerBriefs, concise summaries of cutting-edge research and practical applications across a wide spectrum of fields. Featuring compact volumes of 50 to 125 (150 as a maximum) pages, the series covers a range of contents from professional to academic in the following research areas carried out at Politecnico:

- Aerospace Engineering
- Bioengineering
- Electrical Engineering
- Energy and Nuclear Science and Technology
- Environmental and Infrastructure Engineering
- Industrial Chemistry and Chemical Engineering
- Information Technology
- Management, Economics and Industrial Engineering
- Materials Engineering
- Mathematical Models and Methods in Engineering
- Mechanical Engineering
- Structural Seismic and Geotechnical Engineering
- Built Environment and Construction Engineering
- Physics
- Design and Technologies
- Urban Planning, Design, and Policy

Paolo Rosa · Sergio Terzi
Editors

Circular Approaches for the Ecodesign, Repair and Remanufacturing of Car and Mass Electronics

The CIRC-UITS Project

POLITECNICO
MILANO 1863

Springer

Editors
Paolo Rosa
Department of Management, Economics and Industrial Engineering
Politecnico di Milano
Milan, Italy

Sergio Terzi
Department of Management, Economics and Industrial Engineering
Politecnico di Milano
Milan, Italy

ISSN 2191-530X ISSN 2191-5318 (electronic)
SpringerBriefs in Applied Sciences and Technology
ISSN 2282-2577 ISSN 2282-2585 (electronic)
PoliMI SpringerBriefs
ISBN 978-3-032-04300-9 ISBN 978-3-032-04301-6 (eBook)
https://doi.org/10.1007/978-3-032-04301-6

This work was supported by Politecnico di Milano.

This Springer imprint is published by the registered company Springer Nature Switzerland AG
The registered company address is: Gewerbestrasse 11, 6330 Cham, Switzerland

Contents

Introduction

Paolo Rosa, Sergio Terzi, and Bernd Kospacek

Abstract This chapter aims at clarifying the main research contents and presenting the main objective of the CIRC-UITS project.

1 Main Context

Considering the increasing demand for data gathering, management and analysis, the global economy is becoming even more semiconductor-dependent. Several sectors are embedding within their products lots of sensors, actuators, electronic control units and telecommunication systems allowing a direct interconnection of the product with the world wide web and a real time exchange of information. Among these sectors, mass electronics and automotive [1] sectors are those that, more than others, have seen an exponential adoption of these technologies since many years. However, a strong dependency from semiconductor-based systems is presenting (especially in the last years) several weaknesses and risks that Europe must solve as soon as possible. Among them, the most important issue relates to the lack of European semiconductor companies and the full dependency from extra-EU suppliers (mainly China and Taiwan). This lack brought the European economy towards the current semiconductor crisis who is strongly influencing some of its strategic markets (e.g. automotive) [2]. Another important issue relates to the scarce ability of the European economy to recover strategic components/materials embedded in electronic equipments and exploit them to make new (high value) products. This way,

P. Rosa (✉) · S. Terzi
Department of Management, Economics and Industrial Engineering, Politecnico Di Milano, Milan, Italy
e-mail: paolo1.rosa@polimi.it

S. Terzi
e-mail: sergio.terzi@polimi.it

B. Kospacek
SAT—Austrian Society for Systems Engineering and Automation, Vienna, Austria
e-mail: bernd.kopacek@sat-research.at

P. Rosa and S. Terzi (eds.), *Circular Approaches for the Ecodesign, Repair and Remanufacturing of Car and Mass Electronics*,
PoliMI SpringerBriefs, https://doi.org/10.1007/978-3-032-04301-6_1

high semiconductor-dependent sectors became some of the most environmental-impacting markets worldwide [3]. From this side, End-of-Life Vehicles (ELVs) [4–6] and e-wastes have been renown since many years like two of the most important sources of secondary raw materials. Trying to cope with all these challenges, the European Commission (EC) published (and in some cases is still working on) specific EU strategies/directives for automotive, e-waste (e.g. Digital Product Passport) and, specifically, semiconductors (e.g. European Chips Act). However, trying to make these sectors more sustainable, circular and resilient, it is mandatory to boost both EoL strategies (e.g. sorting, reuse, remanufacturing and recycling) and intra-EU production through innovations and investments. The current international scenario represents a good chance to decouple the European economy from both natural resource depletion (e.g. Critical Raw Materials—CRMs) and dependency from extra-EU supplies of strategic products. In order to better prove what the benefits of a joined circular/resilient use of secondary resources are, the automotive and mass electronics sectors have been identified as the reference for establishing a set of innovative solutions.

2 Main Objectives

To this aim, the CIRC-UITS project demonstrates the improvement to the circularity of automotive and mass electronics sectors by recovering materials from waste products, as well as supporting the reuse & remanufacturing of electronic components into new (high value) products in these sectors (Fig. 1).

In this scenario the main goals of CIRC-UITS are:

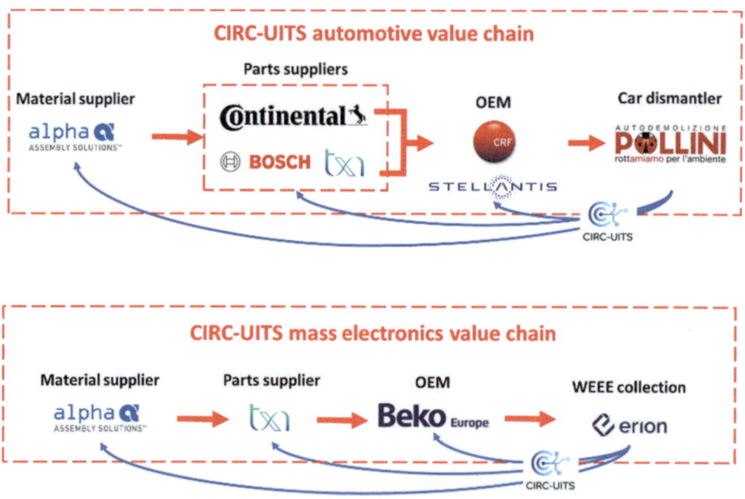

Fig. 1 Overall concept of the CIRC-UITS project

- Unlock full potentials/benefits of circular practices through digital technologies. Sub-intents are: (1) Support the transition towards circular practices through the adoption of different digital technologies, (2) Identify/track/share data about critical/hazardous materials throughout their lifecycle (e.g. materials chemistry, origin, state of health and chain of custody), (3) Define a Product Environmental Footprint and enable the EU Environmental Technology Verification, (4) Link reference sectors with the EU LCA platform, (5) Decrease transaction costs and increase collection rates.
- Increase resource efficiency/independency and reducing the negative environmental footprint of electronics production processes through circular behaviours. Sub-intents are: (1) Increase resource efficiency/independency from imported materials, (2) Reduce the negative environmental footprint of current manufacturing practices, (3) Identify the best EoL scenarios in order to increase reuse, refurbish, remanufacturing and recycling.
- Improve/standardize information/data sharing/exchange among industrial leaders involved in the same and/or similar value chain. Sub-intents are: (1) Demonstrate in practice what are the real benefits coming from the adoption of circular behaviours, (2) Organize/implement dissemination, communication and education actions.
- Demonstrate the benefits coming from Digital Circular Economy through 4 pilots: P1 (Electronic Control Unit—ECU) will support the development of circular ECUs embedded in brake systems; P2 (Tyre Pressure Monitoring Sensor—TPMS) will support the development of circular tyre sensors; P3 (In-Mold Electronics—IME) will support the development of circular IME with fully embedded alternative to PCB based on printed electronics; P4 (Obsolete PCBs) will support the classification and storage of PCBs from different WEEEs.

References

1. https://www.marketsandmarkets.com/pdfdownloadNew.asp?id=162948952
2. https://www.kearney.com/communications-media-technology/article/?/a/europes-urgent-need-to-invest-in-a-leading-edge-semiconductor-ecosystem
3. EU Circular Economy Action Plan (CEAP)
4. https://appsso.eurostat.ec.europa.eu/nui/show.do?dataset=env_waselvt&lang=en
5. Andersen FM, Larsen HV, Skovgaard M, Isoard S (2008) Projection of end-of-life vehicles. Eur Top Cent Resour Waste Manage 3–5
6. Reuter MA, Hudson C, van Schaik A, Heiskanen K, Meskers C, Hagelüken C (2013) Metal recycling: opportunities, limits, infrastructure, a report of the working group on the global metal flows to the international resource panel. UNEP—United Nations Environment Programme

Potential of Digital Tools for Sustainability Reporting

Tobias Hoiten, Maria Fernanda Davila Restrepo, and Alexandra Pehlken

Abstract Sustainability reporting is becoming increasingly challenging for companies—especially in view of the new CSRD. Digital tools are required to present the large amount of data in the complex corporate systems and along the supply chain in a transparent and credible manner. However, using these is sometimes expensive, requires a lot of knowledge and takes time. The corporate carbon footprint is an important instrument in this construct for determining a company's carbon footprint. We have developed a tool with which even small and medium-sized companies can create and visualize this with little effort.

Code available in GitHub.

1 [1]Introduction

Sustainability reporting is challenging for organizations seeking to demonstrate their commitment to environmental, social, and governance (ESG) goals. Stakeholders demand greater transparency and accountability while companies need to compile, analyze and present complex data across their whole supply chain. Manual processes and the use of disparate data systems for reporting in the past made it prone to errors, efficiency and real-time and actionable insights.

The use of digital tools can transform sustainability reports into more accurate and reliable reports. Advanced technologies such as data analytics, artificial intelligence

[1] https://github.com/mafedavila/sustainability_reporting

T. Hoiten (✉) · A. Pehlken
OFFIS Institute for Computer Science, R&D Division Manufacturing, Oldenburg, Germany
e-mail: tobias.hoiten@offis.de

A. Pehlken
e-mail: alexandra.pehlken@offis.de

M. F. D. Restrepo
Carl Von Ossietzky University Oldenburg, Oldenburg, Germany
e-mail: maria.fernanda.davila.restrepo@uni-oldenburg.de

P. Rosa and S. Terzi (eds.), *Circular Approaches for the Ecodesign, Repair and Remanufacturing of Car and Mass Electronics*,
PoliMI SpringerBriefs, https://doi.org/10.1007/978-3-032-04301-6_2

(AI), and cloud computing can streamline the collection, management, and dissemination of sustainability data. Additionally, companies become able to track progress against sustainability targets like the carbon footprint that becomes mandatory in the Corporate Sustainability Reporting Directive (CSRD). Europe-wide and globally recognized frameworks like the CSRD and Global Reporting Initiative (GRI) demand comparability, which digital tools can support.

However, the adoption and effective use of digital tools can encounter barriers: The implementation costs can be high, the complexity of integrating the new systems can be high and there can be a lack of knowledge.

Our past work with focus solely on Product Carbon Footprint (PCF) has shown how these barriers and challenges can be overcome with simple digital tools—especially for small and medium-sized enterprises. This chapter presents a complete sustainability reporting tool compliant with Corporate Sustainability Reporting Directive as mentioned above. It includes two perspectives: Corporate Carbon Footprint (CCF) and Product Carbon Footprint (PCF). A summary of the main contributions analyzed is reported in the following sub-sections [7].

2 State of the Art Sustainability Reporting/Related Work

This section presents a brief literature analysis related to relevant studies describing the state of the art for sustainability reporting. The attention was placed on an interdisciplinary character, aspects from economics as well as from related engineering domains like sustainable engineering and computer science are relevant to better understand the related work. A summary of the main contributions analyzed is reported in the following sub-sections.

Companies in Western Europe and the United States started in the 1970s with sustainability reporting as a tool for publishing environmental and social aspects [18, 19]. After the Brundtland definition of Sustainable Development got popular in 1987 companies started to publish their environmental, social and economic activities in their sustainability reports [31].

Reporting about sustainability information is beneficial for companies in two ways: the first is for building trust and inform the shareholders and a wide array of stakeholders [15] and the second is for increasing credibility ethical behaviour along the supply chain and reducing legal risks [24].

The new Corporate Sustainability Reporting Directive (CSRD) [13] made the shift from “non-financial reporting” to “sustainability report” [1]. The former Non-Financial Reporting Directive (NFRD) enhanced corporate transparency and accountability regarding the ESG (environmental, social, governance) issues while the CSRD is forcing with its European Sustainability Reporting Standards (ESRS) more comparability and reliability. The European Reporting Advisory Group [12] plays a major role in shaping sustainability reporting standards within the European Union supporting also small and medium-sized enterprises in understanding and implementing the ESRS. The main differences to the NFRD are the double materiality analysis, an audit requirement, a larger number of obligatory companies and a clear and comparable reporting format.

2.1 *Role of the Environmental Sustainability in the CSRD*

The role of environmental sustainability in sustainability reporting is becoming increasingly important due to the "European Green Deal" making Circular Economy topics a key strategy of the European Union [13]. The global sustainability trend highlights the problem of highly polluting waste in mature and complex industries [4]. The ESRS force companies in the EU to more data-driven and transparent reports, offering stakeholders clearer insights into a company's environmental performance. Especially the environmental standards from ESRS E1 to E5 contain topics such as climate change, pollution, water and marine resources, biodiversity and ecosystems and resource use and circular economy. For the first time companies are forced to report about their circular economy strategies [12].

In connection with this framework and the comparable reporting format companies need to draw up their corporate carbon footprint (CCF) while small and medium-sized enterprises also receive inquiries about their product carbon footprint (PCF) that precisely records the CO_2 emissions of a product. The PCF measures the total amount of greenhouse gas emissions that are directly and indirectly linked to a product. It takes into account the entire life cycle of the product, from raw material extraction, production and transport to product use and disposal. In our previous work, we showed how easy it is for companies to calculate PCF. In our previous work, we showed how easy it is for companies to calculate the PCF. In this work, we want to go a step further and show a tool that can be used to calculate and visualize the CCF for the entire company. The CCF is a metric to assess a companies contribution to climate change by quantifying its GHG emissions.

For instance, the Ellen MacArthur Foundation [11] advocates for organizations to report on commitments, strategies, risks, opportunities, and impact metrics across all three CE principles: eliminating waste and pollution, circulating products and materials at their highest value, and regenerating nature.

2.2 *Overview of Digital Tools for Sustainability Reporting*

Digital tools play a transformative role in sustainability reporting by improving data accuracy, streamlining processes, and enhancing the accessibility and transparency of environmental, social, and governance (ESG) disclosures. There are different type of tools to draw up a PCF or CCF with different functions such as data collection, visualization, reporting management, blockchain and AI tools.

Tools for data collection and general management tools like SAP Sustainability Control Tower [26] or IBM Envizi [16] can storage the automated data input and monitor them real-time while reporting tools for sustainability reporting like Workiva [30] and Diligent ESG [8] align with global frameworks such as GRI and can format the report. Advanced analytics platforms for data collection and management like SAP Sustainability Control Tower or IBM Envizi, such as those incorporating artificial intelligence (AI) and machine learning such as BrightAI [2], allow organizations to automate the collection and integration of complex ESG data across supply chains and monitoring it in real-time [3]. Blockchain-based tools such as Circularise [5] or Everledger [14] further enhance the credibility of sustainability reporting by enabling traceability and ensuring the immutability of disclosed data, fostering trust among stakeholders [25]. Furthermore, digital dashboards and visualization platforms like Tableau [28] or Microsoft Power BI [21] facilitate the communication of sustainability metrics in an understandable and actionable format, catering to the diverse needs of investors, regulators, and the public [20].

However, there are barriers to adoption these tools for companies such as the difficulty of technological integration, high initial costs and lack of expertise. Especially small and medium-sized enterprises must address challenges related to cost, complexity, and standardization while leveraging opportunities for innovation and broader adoption to maximize their benefits.

That's why we developed a sustainability reporting tool for the CCF which contains data collection, general management and a digital dashboard with an easy way of adoption, no initial costs and less knowledge to implement (Table 1).

3 Methods

3.1 Emissions Calculation

In this section, the emissions calculation of the CCF and the PCF are described. Inspired by Davila et al. (Davila et al., in press, 2025) the digital tool can calculate the CCF in addition to the PCF. The necessary KPIs for CCF are shown in Table 2 and PCF in Table 3.

To calculate emissions, we employ the primary Eq. (1), where E represents the calculated emissions, A denotes the activity rate, EF signifies the emission factor, and ER indicates the emission reduction efficiency.

$$E = A \times EF \times \left(\frac{1 - ER}{100}\right) \tag{1}$$

For the calculation of material emissions, we utilize the sum of the mass of material used per component (Mmaterial), yielding Eq. (2). To determine transport emissions, we use the sum of tons per kilometer of transported components (Mtransported),

Table 1 Input data tables needed for CCF calculation

Table name	Description
Companies	Contains details about the organization, including the company ID, name, and location
Factories	Stores information about individual factories within the company, including their ID, name, and location
Materials	Tracks material usage, such as types, quantities, transportation methods, and distances. Many manufacturing companies already have this information stored in ERP (Enterprise Resource Planning) software [27]. ERP systems are designed to manage and integrate core business processes, including supply chain management, inventory tracking, production planning, and resource allocation
Energy usage	Records energy consumption, including energy source types, timestamps, and quantities in kilowatt-hours. This is typically stored in an Energy Management System (EMS) [17], which is a software platform that monitors, controls, and optimizes energy consumption within an organization to improve efficiency and reduce costs
Water usage	Contains data on water consumption, including total liters used and timestamps. Sometimes also stored in the EMS
Waste management	Logs waste generation data, such as waste types, amounts, and timestamps
Emission factors	Stores emission factors for materials and energy sources, facilitating automated calculations. We use materials databases, such as Probas [29], Ecoinvent [10] or OpenCO2Net [23]
Energy emission factors	Includes emission factors specific to energy sources, such as emissions per kilowatt-hour. This depends from the energy provider and the energy mix used by the company

Table 2 Calculated KPIs for the CCF in the sustainability report

View name	Description
Total waste generated	Aggregates the total waste by factory and month, providing insights into waste management efficiency
Total water usage	Summarizes water consumption for each factory and time period, highlighting areas of high-water usage
Energy intensity per product	Calculates energy consumption normalized by product output, measuring production efficiency
Material emissions	Quantifies emissions associated with material usage, enabling evaluation of supply chain sustainability
Transport emissions	Evaluates emissions related to material transportation, identifying opportunities for logistical optimization
Energy emissions	Summarizes emissions from energy usage, supporting energy efficiency strategies
Total emissions	Combines all emission sources to provide a comprehensive carbon footprint for each factory

Table 3 Tables with the input data for PCF calculation

Table name	Description
Order	Captures production details, including product ID, factory, production start and end dates, and hall location. This is typically stored in the Manufacturing Execution System (MES), a software system that manages, monitors, and controls production processes in real-time to ensure efficient and optimized manufacturing operations
Energy	Records energy consumption data at the machine level
Energy sources	Provides detailed information about energy types, whether they are conventional or renewable sources, and corresponding emission factors
Machines	Includes machine-specific data, including nominal power, and associated processes and products
Workplan	Details the production plan of a product, linking its product serial with the machine information
Materials	Stores information on materials used per product, including emission factors, data sources, and quantities
Components	Details the product components, suppliers, and material composition
Suppliers	Contains supplier-specific data, such as distances from the factory and transportation used by them

resulting in Eq. (3). The emission factor EFtransport was assumed to be equivalent to that of a heavy goods vehicle (according to Probas [P24]). For the calculation of energy emissions, we require the energy consumption per component and product (Eprocess), leading to Eq. (4). The energy emission factors EFenergy depend on the energy mix, which represents the percentage of energy generated from each source at the time of production. Finally, the total emissions of the product (Eproduct) are calculated as the sum of material, transport, and energy emissions, as shown in Eq. (5) [7].

$$Ematerial = \sum Mmaterial \times EFmaterial \tag{2}$$

$$Etransport = \sum Mtransported \times EFtransport \tag{3}$$

$$Eenergy = \sum Eprocess \times EFenergy \tag{4}$$

$$Eproduct = Ematerial + Etransport + Eenergy \tag{5}$$

3.2 Digital Tool

The tool is composed of four independent services: (i) a database for the Corporate Carbon Footprint (CCF) data, (ii) a database for the Product Carbon Footprint (PCF) data, (iii) a dashboard to visualize the results, and (iv) a web-based platform that allows users to interact with the tool.

The tool was developed using a microservices architecture [22], implemented with Docker [9] Compose. Microservices architecture is a design approach in which an application is structured as a collection of loosely coupled, independently deployable services. Each service corresponds to a specific functionality, such as database management, data visualization, or user interaction. This modular design provides several advantages, including improved scalability, fault isolation, and flexibility in technology choices. For example, the database services and the web-based platform can be developed using different programming languages or frameworks, allowing each service to be optimized for its specific purpose.

Docker Compose is a tool that simplifies the management of microservices applications. By defining the tool's architecture in a docker-compose.yml file, all services can be configured, built, and launched simultaneously. The main advantages of using Docker Compose include:

- Ease of Deployment: Developers can start the entire application with a single command, ensuring consistency across environments.
- Isolation: Each service runs in its container, preventing interference between services.
- Portability: Docker Compose configurations make it easy to replicate the application on any machine that supports Docker, ensuring cross-platform compatibility.

Our tool is open source, meaning that anyone can access its code on GitHub[1]. Users can follow the instructions provided to install Docker and build the application on their local computer. Upon installation, the example use case described in Sect. 3 is automatically set up. This includes synthetic data from a Truck Trailer company in northern Germany, which serves as a demonstration dataset. Users can replace this example data with their own company's resource consumption data, and the tool will calculate both the CCF and the PCF.

3.3 Database for the CCF Data

The database for the CCF data is built using a relational database schema [6]. Relational databases organize data into tables with rows and columns, allowing efficient querying, updating, and linking of information. This approach ensures data consistency, integrity, and scalability, making it ideal for managing complex datasets such as resource consumption and emissions data.

The schema of the database is designed to capture all relevant data required for Corporate Carbon Footprint (CCF) calculations. It is divided into several interrelated tables, shown in Table 1.

These tables represent the input data collected from the company, which is structured and pre-processed by the digital tool to calculate the main components of the sustainability report. The calculated views, which summarize and analyze the data are shown in Table 2.

These calculated views form the core outputs of the sustainability report and are displayed on the dashboard. They enable users to monitor, analyze, and improve their resource consumption and emission metrics effectively. The database's structure ensures that users can replace synthetic data with real-world data, allowing the tool to adapt to diverse organizational needs.

3.4 Database for the PCF Data

The database for the Product Carbon Footprint (PCF) data is specifically designed to calculate emissions at the product level, utilizing detailed data from production plans, machines, and energy consumption. It is built using a relational schema and consists of several interconnected tables to ensure accuracy and traceability in emissions calculations, shown in Table 3.

The input data from these tables is pre-processed and structured by the digital tool to calculate three key emissions components for each product: energy, materials, and transportation emissions. Using these components, the tool calculates the Total Product Emissions, which provide a comprehensive carbon footprint for each product. The calculated views displayed on the dashboard are shown in.

View name	Description
Energy consumption	Summarizes energy consumption by product
Energy emissions	Calculates emissions from energy usage for each product
Material emission	Quantifies emissions from material usage at the product level
Transport emission	Evaluates transportation emissions for product components
Product footprint	Combines energy, material, and transport emissions to provide the total carbon footprint for each product

4 Use Case

The tool was tested as a pilot using synthetic data from a truck trailer company in northern Germany. Figure 1 shows the platform's interface, which features a menu on the left side allowing users to navigate between corporate and product reports/dashboards.

The platform is built using Angular, a popular framework for building dynamic, web-based applications. Angular allows for the creation of a highly customizable and responsive user interface, ensuring that users can interact seamlessly with the tool. Additionally, the platform can be adapted by any user based on the open-source code available on GitHub.

4.1 Corporate Carbon Footprint Results

This subsection shows the key outputs of the Corporate Carbon Footprint calculations. Figure 2 displays the results for material consumption. The platform automatically generates a bar plot that shows material consumption in mass by material type (e.g., Steel, Aluminum). Alongside this, a donut chart visualizes material emissions by material type. The example data highlights that, while electronics do not have the highest mass, they are the largest contributor to material emissions. This visualization allows users to identify key areas for emission reduction and material efficiency improvements.

In addition, transportation emissions results can be displayed as well, which are calculated based on the distance to each supplier, the means of transportation used, and the weight of goods transported. The results are presented as both the mass of CO_2 equivalent emissions and the percentage contribution of each supplier. This information can be used to identify suppliers with the highest transportation emissions and prioritize efforts to optimize logistics or switch to more sustainable transportation methods.

Figure 3 presents the energy emissions, calculated first in megawatt-hours (MWh) with energy sources classified into conventional and renewable categories. Using

Fig. 1 Home page of the platform

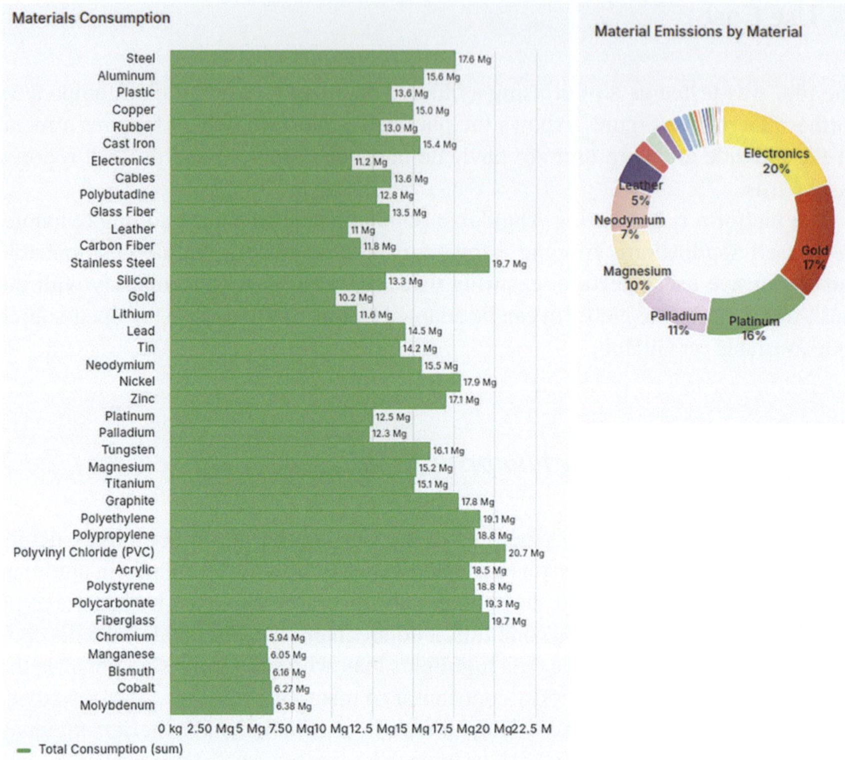

Fig. 2 Results of the material emissions and consumption inside the CCF module of the dashboard

predefined conversion factors for each energy source, the results show the total energy emissions per source, enabling a clear comparison of their environmental impact. Additionally, the figure highlights the energy intensity, which measures energy efficiency by showing the energy consumption per unit produced (kWh/unit) and per revenue generated (kWh/revenue). This metric provides valuable insights into the energy efficiency of the production process, helping organizations identify areas for improvement.

Figure 4 presents waste generation categorized by factory and material groups, such as metals and plastics, providing actionable insights for resource management and waste reduction. In addition, it is possible to visualize the water consumption across different factories within the company.

Figure 5 shows the final total emissions per year, for the last three years. It is first presented as the total emissions and then by material, transport and energy emissions.

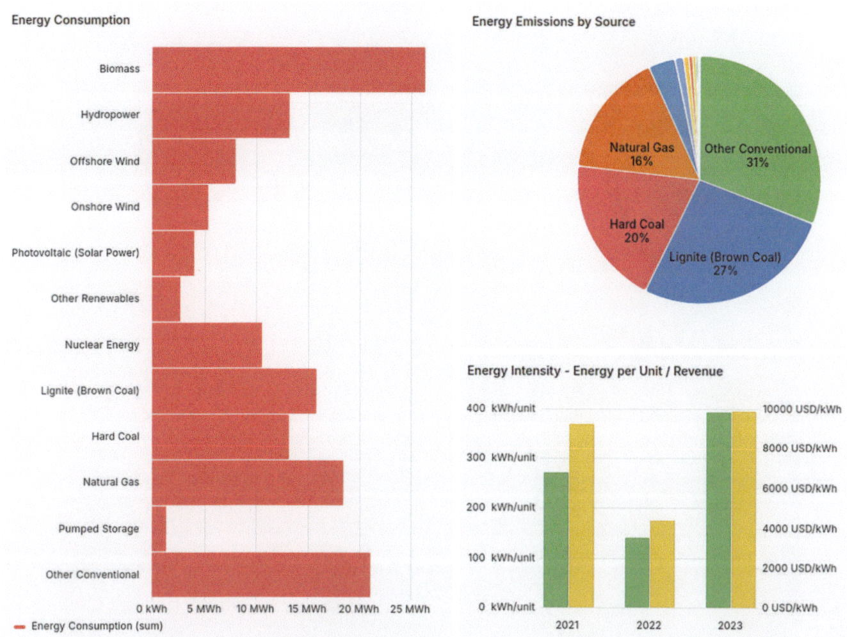

Fig. 3 Results of the energy consumption and emissions per energy source, including energy intensity metrics per unit produced and per revenue

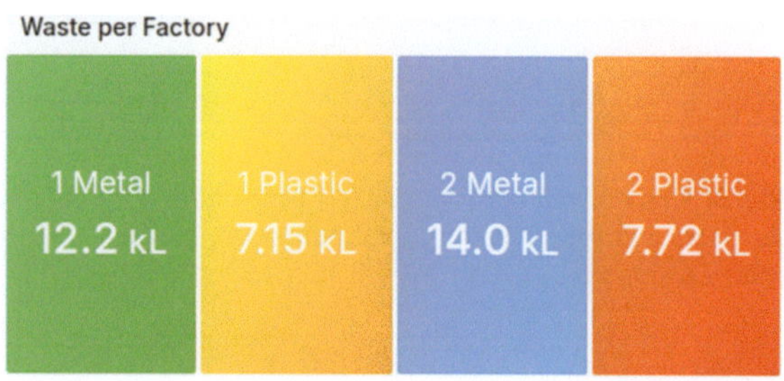

Fig. 4 Waste generation per factory and material group

Fig. 5 Corporate carbon footprint (CCF) results over the last three years, with columns representing the years and rows detailing total emissions, material emissions, transportation emissions, and energy emissions

5 Conclusion

The digital dashboard, as presented here, helps large as well as small and medium-sized companies to comply with the sustainability regulations regarding the CSRD and ISO 14067 standards. Both the calculation and visualization of the CFP and the CCF are necessary for companies to comply with the ESRS. While other tools are usually confusing, costly and require a lot of knowledge, with the help of our open-source design tool you can quickly and easily create a free tool to calculate and display the CFP and CCF with fewer resources. This not only improves the accuracy of sustainability reports but also supports the larger goal of reducing climate impact as part of the European Green Deal. The user-friendly dashboard shows emissions data clearly, helping companies spot areas to improve in their processes.

References

1. Adams CA (2017) Conceptualising the contemporary corporate value creation process. Account Audit Account J 30(4):906–931. https://doi.org/10.1108/AAAJ-04-2016-2529
2. Bright AI (2024) BrightAI platform. Retrieved from https://www.bright.ai/
3. Burritt RL, Christ KL (2021) Sustainability accounting research and the contribution of digital technologies. Account Forum 45(4):331–362
4. Cardamone GF, Ardolino F, Arena U (2021) How to enhance the environmental sustainability of WEEE plastics management: an LCA study. Waste Manage 135:347–359. https://doi.org/10.1016/j.wasman.2021.09.021
5. Circularise (2024) Product traceability platform for supply chain compliance. Retrieved from https://www.circularise.com/
6. Date C (2005) Database in depth: relational theory for practitioners. O'Reilly. Retrieved 19 Dec 2024
7. Davila R, Maria F, Hoiten T, Sander P, Woltering N, Pehlken A (2024) Container-based microservices application for product carbon footprint calculation in manufacturing companies. In: EnviroInfo 2024, pp 75–85. Business Intelligence for Green Decision-Making, Cairo, Egypt, Gesellschaft für Informatik e.V, Bonn EISSN: 2944-7682. https://doi.org/10.18420/env2024_07
8. Diligent (2024) ESG solution-based software. Retrieved 19 Dec 2024 from https://www.diligent.com/solutions/modern-esg
9. Docker (2024) Docker: accelerated container application development. https://docs.docker.com/reference/
10. Ecoinvent (2024) Ecoinvent database. Retrieved 19 Dec 2024 from https://ecoinvent.org/
11. Ellen MacArthur Foundation (2013) Towards the circular economy: economic and business rationale for an accelerated transition. Ellen MacArthur Foundation
12. European Financial Reporting Advisory Group (EFRAG) (2023) Sustainability Reporting. Retrieved on 14 Nov 2024 from https://www.efrag.org
13. European Commission (2021) Proposal for a DIRECTIVE OF THE EUROPEAN PARLIAMENT AND OF THE COUNCIL amending Directive 2013/34/EU, Directive 2004/109/EC, Directive 2006/43/EC and Regulation (EU) No 537/2014, as regards corporate sustainability reporting, COM(2021) 189 final. European Commission, Brussels. https://eur-lex.europa.eu/legal-content/EN/TXT/PDF/?uri=CELEX:52021PC0189&from=EN
14. Everledger (2024) Everledger platform. Retrieved 19 Dec 2024 from https://everledger.io/our-platform/
15. Freeman RE (1984) Strategic management: a stakeholder approach. Pitman, Boston
16. IBM (2024) Envizi. Retrieved 19 Dec 2024 from https://www.ibm.com/de-de/products/envizi
17. ISO (2018) ISO 500001 Energy management. Retrieved 19 Dec 2024, from http://www.iso.org/iso/home/standards/management-standards/iso50001.htm
18. Junior RM, Best PJ, Cotter J (2014) Sustainability reporting and assurance: A historical analysis on a world-wide phenomenon. J Bus Ethics 120(1):1–11. https://doi.org/10.1007/s10551-013-1637-y
19. Kolk A (1999) Evaluating corporate environmental reporting. Business Strategy and the Environment 8(4):225–237. https://doi.org/10.1002/(SICI)1099-0836(199907/08)8:43.0.CO;2-4
20. KPMG (2022) The future of ESG data and reporting. Retrieved on 14 Nov 2024 from kpmg.com
21. Microsoft (2024) Power BI. Retrieved from https://www.microsoft.com/de-de/power-platform/products/power-bi?market=de
22. Newman S (2015) Building microservices. O'Reilly Media. Retrieved 19 Dec 2024
23. OpenCO2Net (2024) OpenCO2Net emission factors. Retrieved from https://www.openco2.net/en/search-emission-factors?sortBy=updatedYear&sortOrder=desc&page=1
24. Paun D (2018) Corporate sustainability reporting: An innovative tool for the greater good of all. Bus Horiz 61(6):925–935. https://doi.org/10.1016/J.BUSHOR.2018.07.012
25. Saberi S, Kouhizadeh M, Sarkis J, Shen L (2019) Blockchain technology and its relationships to sustainable supply chain management. Int J Prod Res 57(7):2117–2135

26. SAP (2024) SAP Sustainability control tower. Retrieved from https://www.sap.com/products/scm/sustainability-control-tower.html
27. Sheilds MG (2005) E-Business and ERP: rapid implementation and project planning. Wiley and Sons, Inc. Retrieved 19 Dec 2024
28. Tableau (2024) Analytics platform. Retrieved from https://www.tableau.com/de-de
29. Umweltbundesamt (2024) ProBas. Retrieved from https://www.probas.umweltbundesamt.de
30. Workiva (2024) Software für ESG, GRC und Finanzberichterstattung. Retrieved from https://www.workiva.com/de/
31. World Commission on Environment and Development (1987) Report of the World Commission on Environment and Development, vol 4, pp 17–25. Our Common Future (The Brundtland Report). https://doi.org/10.1080/07488008808408783

The Development of a Circularity Toolbox for Sustainable Product Design

Lisa Dawel, Mattia Calabresi, Francesca Lazzari, Aleksandr Bystrov, Ole Meyer, Felix Schmedes, Antoinette van Schaik, Julien van Damme, Markus A. Reuter, and Alexandra Pehlken

Abstract This chapter provides an overview of the digital tools that support sustainable product design in the CIRC-UITS project via the circularity toolbox. The toolbox is designed in such a way to accommodate for different needs and is flexible by integrating different tools with distinct functionalities. In this chapter, case studies highlight real-life application of the tools. The functionalities and tools are presented

L. Dawel (✉) · A. Bystrov · O. Meyer · F. Schmedes · A. Pehlken
OFFIS Institute for Computer Science, R&D Division Manufacturing, Oldenburg, Germany
e-mail: lisa.dawel@offis.de

A. Bystrov
e-mail: aleksandr.bystrov@offis.de

O. Meyer
e-mail: ole.meyer@offis.de

F. Schmedes
e-mail: felix.schmedes@offis.de

A. Pehlken
e-mail: alexandra.pehlken@offis.de

M. Calabresi · F. Lazzari
TXT E-TECH S.R.L., Cologno Monzese, MI, Italy
e-mail: mattia.calabresi@txtgroup.com

F. Lazzari
e-mail: francesca.lazzari@txtgroup.com

A. van Schaik · M. A. Reuter
Material Recycling and Sustainability (MARAS) B.V., Den Haag, The Netherlands
e-mail: a.vanschaik@maras-bv.nl

M. A. Reuter
e-mail: markusandreasreuter@icloud.com

J. van Damme
Department of Management, Economics and Industrial Engineering, Politecnico Di Milano, Milan, Italy
e-mail: julien.vandamme@polimi.it

P. Rosa and S. Terzi (eds.), *Circular Approaches for the Ecodesign, Repair and Remanufacturing of Car and Mass Electronics*,
PoliMI SpringerBriefs, https://doi.org/10.1007/978-3-032-04301-6_3

in the order of the lifetime phases of physical products, showcasing the unique value that they can bring to each phase.

1 The Need for Digital Tools in Assessing Product Circularity

In addition to ever-increasing energy consumption, the demand for material resources for manufacturing electronic components is coming to the forefront. This challenge is reinforced by two aspects: (a) short lifetime and high embodied energy and (b) increasing material demand. As ICT products have short life cycles, not only the energy requirement in the operation phase is relevant. The so-called embedded energy, i.e., the energy that is required for the manufacture, transport, storage, and disposal of the components, can also make up a considerable proportion of the total energy requirement [1]. On the other hand, this also results in an increasing demand of raw materials for these applications. This aspect is becoming increasingly important as the material intensity of ICT products increases [2]. The improvement of circularity of the automotive and mass electronics industries, due to the growing scarcity of raw material and semiconductors, is a main challenge. This implies both environmental (depletion of resources) and business/political risks (dependence on import from extra European countries) [3].

The improvement of circularity in the above-mentioned sectors implies a set of sub-challenges and needs to be taken into account:

- **Remanufacturing and repairability** (assessment of feasibility).
- **Re-use** both of as-is and repaired/remanufactured components (identifying barriers and basic standards).
- **Eco-design** in the Beginning of Life (BoL) phase of the product
- **Closing the circle** in value chain by connecting Beginning of Life and End of Life (EoL) operators.
- Selection of **improved KPIs** for measuring sustainability and circularity taking into account manufacturing, repairability, recyclability and re-use practices.

2 Architecture of a Circularity Toolbox

In this context, a circularity toolbox encompasses a set of services aimed at supporting BoL and EoL management in the strategic sectors of mass electronics and automotive with the objective of recovering materials from waste products and remanufacturing, repairability, re-use electronic components into new high-value products.

The toolbox of the EU Project CIRC-UITS was designed based on the analysis of the needs of these two industries, which are represented within CIRC-UITS by 4 pilots (Pilot 1 "ESP control units' eco-design process"; Pilot 2 "Tire pressure monitoring sensors' eco-design process"; Pilot 3 "In-mold electronics' eco-design process"; Pilot 4 "Printed circuit boards' sorting process").

A thorough analysis was carried out to define the "as-is" situation and the "to-be" perspective of each pilot, thus assessing the current status of the selected industrial processes and the consequent needs in terms of technologies to be developed and performance indicators to reach the desired "as-is" perspective improved in sustainability and circularity.

This analysis served as baseline for the architecture of the toolbox (see Fig. 1).

The toolbox encompasses five main modules which are designed to support the actors of the automotive and mass electronics sectors to tackle the circularity challenge and improve their sustainability in terms of processes and final products. As depicted in Fig. 1, the major five modules of the CIRC-UITS toolbox are composed of individual back-end functionalities and front-end GUIs. These modules constitute the major assets of the project toolbox:

- **Simulation & Digital Twin module**: the simulation tool is an instrument that can help the user to design and develop more sustainable and circular products based on specified KPIs.
- **Distributed Advisory services with AI-based functionalities**: its objective is twofold—it can help pilots to optimize their product design, by exploiting different analyses of the existing product to formulate new designs and, on the other hand, it aims to help manage the recycling of a product's components at the end of its life cycle. Since the involved toolboxes provide AI-based advisory functionalities, the results of which will be shared with the rest of the platform, the advisory services have to be considered as a "distributed" component. Advisory is presented in detail in the book chapter on Sustainability and Circularity Assessment & Advisory Methodologies and Services.
- **LCS&CA module**: this is a central component aimed at assessing the environmental, economic, social, and circular impacts of current and new products designed in CIRC-UITS. The assessment will not only consider different design options, but will also provide KPIs, decisions and "End-of-Life" alternatives in order to define the most optimal solutions for repair versus recycling and EoL circularity of different products and repair scenarios. This module is presented in detail in the book chapter on Sustainability and Circularity Assessment & Advisory Methodologies and Services.
- **Advanced HMIs module**: it allows the interaction between users and the CIRC-UITS digital toolbox functionalities in specific contexts in which a "traditional" GUI is not enough by implementing complex visualizations that leverage mobile-specific interfaces and augmented/virtual reality procedures.
- **Marketplace module**: dedicated secure tool where users can sell/purchase objects and components to facilitate the reuse and circularity of semiconductors and strategic materials. The Marketplace tool leverages innovative technologies such as agent-based contract negotiation strategies and blockchain-based smart contract agreements to support and improve asset selling/purchasing.

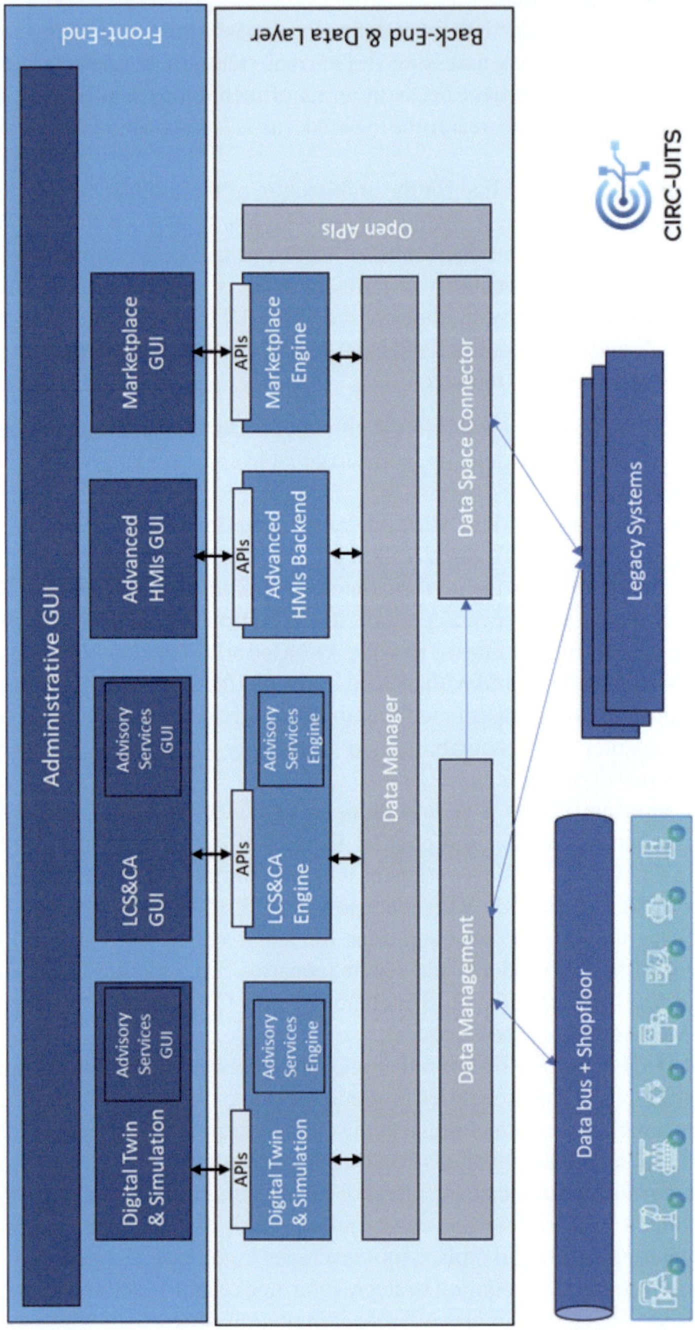

Fig. 1 CIRC-UITS circularity toolbox architecture

Of particular note is the **data sharing** within the toolbox, which takes place via the **Data Layer**. As revealed during the preliminary analysis and the discussions with the end-users (project pilots), companies and other important stakeholders along the value chain are often reluctant to share their process/production data. On one hand, this protects their security; on the other hand, this hinders the improvement of the circularity of the overall industry: the actors involved need specific information about components and materials in order to carry out their own sustainability and circularity assessments, thus improving their performances.

In this regard, a system for secure data sharing among the value chain actors was included in its Circularity toolbox which revolves around the Data Space technology. The Data Space is a novel technology that allows secure data exchange, ensuring that the data owner retains its ownership over their data and is able to set up a set of policies to handle access rights and data management.

Finally, the Data Layer offers a unified, secure, trusted and traceable data storage solution to ingest outside data (shop floor data sources, local data buses, and companies' legacy systems) into the toolbox to favour the flow of information within the rest of the toolbox (see Fig. 2 on page 7).

As previously mentioned, the toolbox and its main modules tackle specific sustainability challenges, to which a set of technologies has been dedicated to support the achievement of higher circularity performances in the target sectors. These challenges serve as use cases that are identified based the product life-cycle phases; they can be summarised as: Beginning-of-Life (BoL including Eco-Design), Mid-of-Life (MoD), Re-use, Repair and End-of Life (EoL -including Material Recovery).

The use cases and the dedicated technologies are described in the following subchapters.

3 Addressing the Whole Lifecycle: Digital Twin and Eco Design

In the beginning of a product's life cycle, the product will be designed and -based on this design- will be produced afterwards. The design phase represents a key phase in the products life cycle since it heavily influences all other following phases. With the eco-design principle, companies are aiming at designing their product in such a way that the resource consumption not only during production is minimized but over the whole life cycle. This includes design for repair or design for recycling.

There are many contradictory aims in eco design for which a trade-off must be made. For that a digital tool is being developed, named the design digital twin that is included in the toolbox. It assesses the sustainability and circularity of a product throughout its life cycle. The tool will help in the following ways:

- It will provide accurate data on the environmental impacts of a product during its design, production, repair and end-of-life phases, by using models that mimic the performance of the product's processing by the various stakeholders.

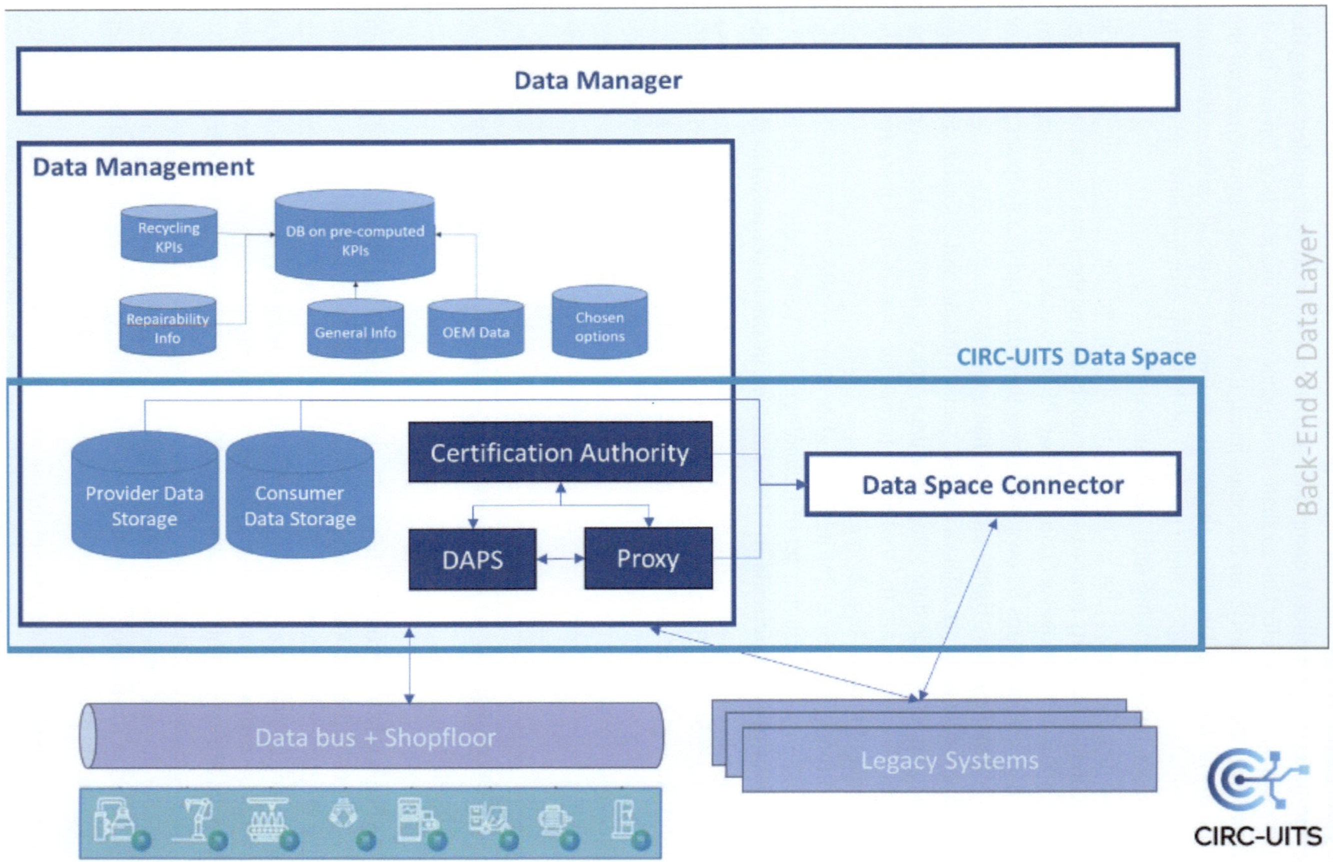

Fig. 2 CIRC-UITS data layer module architecture

- It will support the life cycle engineering (LCE) of the product, by using simulated data to evaluate the environmental aspects of the product design and development as also described in the chapter on Sustainability and Circularity Assessment & Advisory Methodologies and Services.
- It shows the repairability potential of different product designs with a break-down of the product-component-structure in order to find optimization potential for design and EoL processing.
- It visualizes the criticality of the materials of the product with a break-down of the product-component-structure in order to find the most critical product components in view also of recovery to recycle back into the same product. See also below.
- It will allow the user to set sustainability requirements for their product and will show the requirement fulfilment for their design.
- It will support the advisory service of the toolbox, by allowing the user to create and compare different scenarios of product design and development, and to select the best alternatives in terms of sustainability and circularity.

The main functionalities of the digital twin and simulation module are described in the following:

- The DT & Simulation module generates new knowledge from existing data sources, such as databases, which will be combined with the LCA data provided by the GRETA tool, as described in the chapter on Sustainability and Circularity Assessment & Advisory Methodologies and Services.
- By evaluating the influence of using different parameters during the design phase, the users can compare and choose the most suitable parameters from their functional point of view, as well as for their circularity potential. The DT & Simulation module allows the user to simulate the performance of various KPIs based on the selection of different parameters during the design of a product.
- The DT & Simulation module will generate scenarios for repairing which take into consideration all the data and parameters gathered to generate models of the products thus promoting the circular economy.
- It provides a breakdown of the criticality of the product materials both in terms of the calculation influencing factors as well as the product-component-structure.
- It provides a fulfilment analysis of numerical sustainability KPIs that the user can set.

The simulation tool is therefore a powerful instrument that can help the user to design and develop more sustainable and circular products, by using digital twin technology. It is based on the repair and recycling KPIs that are defined from the companies, as seen in the chapter on eco-design of ECUs. This digital twin helps to assess the best environmental performance and is highly efficient in assessing the circularity success for various products. With the integration of AI methods, like machine learning, the best performance of the products could be achieved.

Part of the design for sustainability is also the long-term view. Since we live in a world with finite resources, securing the long-term availability of resources necessary for the production of the product is of strategic interest. Thus, knowing which materials used are critical and acting on it can benefit a company in the long term enormously. Having this knowledge enables companies e.g. to reengineer their own supply chain in such a way to use old products as supply of their critical materials and/or repair parts. For this, detailed knowledge is necessary which can be gained in the critical raw materials dashboard as part of the digital twin:

At least since the semiconductor shortage during the COVID-19 pandemic it became apparent to many that Europe is dependent on outside sources for many components of modern technologies. While the semiconductor shortage was fuelled by a limited supply of components, this dependency runs deeper than that. Even down to the raw materials of our products, the EU is dependent on mining and processing in third party countries [4].

While this is not ideal the real problem arises with geopolitical conflicts and crisis like the war in Ukraine or tensions with China threaten the security of the EUs raw material supply. In the attempt to lower this risk, the EU enacted the Critical Raw Materials Act (CRMA) [4], which aims to increase the production of critical materials in the EU, as well as the reuse and recycling of such materials. Every three years the EU also publishes a report on the most important materials and their current risk of supply disruption, as declared in the CRMA.

While monitoring the EU wide level of supply is important for future EU policies and supply security, a currently overlooked aspect of this monitoring goes down to the companies and their products [5]. Product designers should be informed of supply risks of materials inside the product they are designing to mitigate future supply disruptions of their product. These can seriously impact purchasing prices and in turn profits for the company.

To mitigate this risk and inform designers is the goal of the Critical Raw Materials database and dashboard that is developed at OFFIS. It enables product designers to unravel the reasons specific materials are deemed as critical if they wish to. The included database also enables further processing of the underlying data about supply risk to ultimately determine which components of the product are affected by potential material supply interruptions. Ultimately enabling the tool to highlight which components benefit the most by increased reusability and repairability.

Summarizing, by making the trade-offs visible that need to be addressed in eco-design, the design digital twin enables the users to make sustainable decisions and to future-proof a product design. It supports the user navigating all the options and design combinations.

4 Mid-To-End-Of-Life: Data Sharing, Sustainability and Circularity

When making a purchase decision, the consumer pays attention to the characteristics of the product, such as the characteristics that must be specified, as in the reporting chapter. From the consumer's point of view, the begin-of-life or mid-of-life phase of a product's life cycle is the most prominent. However, for the company producing the product, this is not the case.

When the companies follow the traditional business model of producing and selling the product, this is the phase of the life cycle where the least amount of information is available to the producer and where the producer has no control. When using product- as-a-service approaches, this information deficit for the producer is smaller. When a problem with the product arises, a decision has to be made if the product can be repaired, if parts of the product can be reused or if the product will undergo further disassembly to extract precious materials.

At the end of a product lifecycle, several processes take place towards efficient and effective disposal of embedded components. This route, however, does not take into account a lot of potential for reuse, especially when dealing with products that are being discarded due to a manufacturing defect or mechanical fault that does not involve the electrical and electronical parts of the appliance (e.g. PCB boards). In such a scenario, the recovery of PCB components represents a valuable approach to reduce the amount of e-waste generated by dismantling facilities, while allowing products manufacturers to re-introduce certified used components back into their production loop to meet the ambitious European goals and KPIs in terms of recovered materials to be re-used in the manufacturing process. In addition, valuable feedback can be gathered from the whole reuse process, leading to a better design for reuse that involves the manufacturing process from the beginning with reuse in mind. This is further discussed in the chapter on Sustainability and Circularity Assessment & Advisory Methodologies and Services.

To this extent we are addressing the needs expressed by End of Life (EoL) actors, such as dismantling plants, shredders, recyclers and smelters, and Begin-of-Life (BoL) actors such as appliance producers and component manufacturers. The major challenges that have been identified in the context of components reuse, are the following:

- Lack of connection/communication between BoL and EoL: often recycling plants come in contact with components in good conditions that meet all the criteria to be salvaged from shredding and reused to build new appliances. The problem is the lack of a trusted and secure way to forward this information to appliance manufacturers, notifying them about the presence of such PCBs. On the other hand, appliance producers don't have visibility on the component availability in recycling plants, as there is no way for them to specify the criteria to be checked in order to establish whether a component is suitable for recycling or not. In short, there is no way for BoL and EoL to exchange valuable knowledge and PCB components and parts.
- Lack of incentive from dismantling facilities to deviate from their current shredding workflow. This happens as a result of the fact that this kind of labour-intensive processes always generate very narrow margins, reducing the willingness to dedicate time and effort to deviate from their well-established workflow and, instead, implementing a new reuse process that includes extracting and testing the waste

components, finding an interested manufacturer, signing a suitable agreement and arrange payment and shipping for the selected PCB(s).

In order to better understand the extent to which these challenges impact the targeted markets and actors, domain experts from different backgrounds have been involved in the preliminary analysis and subsequent design and validation of the proposed solution.

The identified solution leads to the identification of the reuse use-case which leverages on three major components of the Circularity Toolbox:

The Marketplace module: centralized solution that enables EoL actors such as recycling plants, WEEE treatment plants, consortia, and BoL actors including appliance manufacturers in the consumer electronics field to come in contact in the scope of objects/components reuse for re-manufacturing. The Marketplace tool allows BoL actors to retrieve used components related to consumer electronic appliances from recycling facilities to be re-included in the manufacturing loop, reducing the impact of EoL recycling activities and enabling circularity among the whole value chain. On the other hand, the Marketplace module enables EoL recycling plants to reduce the amount of component designated to smelting facilities, boosting reuse in a profitable way. In particular, the main objectives of the Marketplace are (i) to allow recycling facilities and WEEE treatment plants to reduce the number of components devoted to shredding and smelting facilities, (ii) to enable EoL actors to meet the ambitious European targets for the recycling of critical and precious materials as a consequence of the usage of second-hand electronic components coming from agreements with the EoL and (iii) to guarantee a more resilient supply of PCBs as spare parts coming from EoL products, therefore creating a new supply chain for the reintegration of PCBs and components and improve the procurement of supplier's components within EU for the mass-electronics sector.

The Dataspace: composed by the Data Management and Data Space Connector sub-modules of the Data Layer (see Fig. 3). This macro-component enables sharing of information between BoL and EoL, in a trusted, secure and traceable manner. It is exploited by manufacturers to share dismantling information related to appliances for which there is interest in recovering internal PCBs. Such information is shared with dismantling plants by means of a series of permissions and rules that can be set by the BoL actor in order to retrain full control over business-critical assets, while also enabling data sharing with recycling facilities in a controlled way, providing insightful details on how to efficiently disassemble appliances. Recyclers, on the other hand, can access such valuable information directly from a trusted source, ensuring information are relevant and can be applied effectively in the dismantling process by exploiting efficient disassembly procedures that reduce the time to extract PCB boards, thus creating a first incentive towards investing in this innovative approach that deviates from the usual shredding workflow.

The Advanced HMIs module: assisting recycling operators in the dismantling activities of WEEE appliances, facilitating the recovery of components and PCBs for reuse. Its main objective is to provide recycling operators with a complete, readable

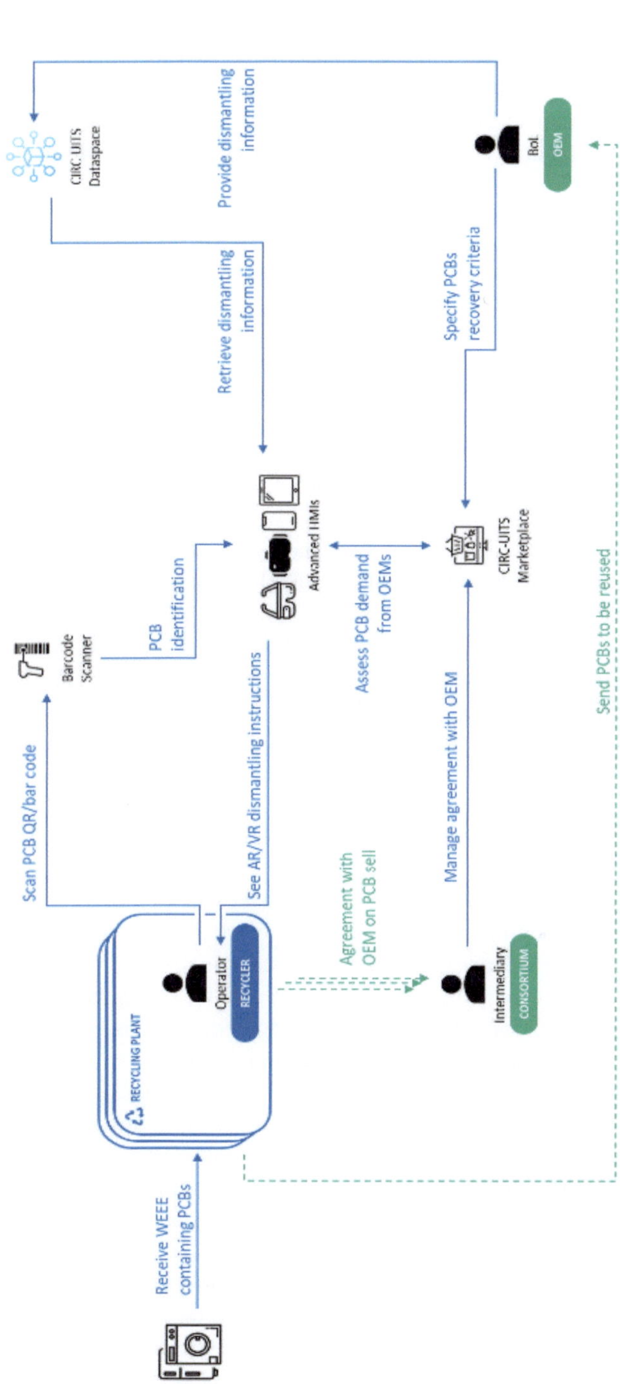

Fig. 3 The interaction diagram highlighting synergies between modules of the CIRC-UITS toolbox, exploited for the PCBs reuse and CRMs recovery use-case

and easy to assess advanced interface with the aim of providing support to perform dismantling procedures of electronic appliances. The Advanced HMIs component is designed to be accessible by the recycling operator in mobility, minimizing the effort required to visualize information and maximizing the flexibility and ease of use. To this aim, the module is offered in various configurations to be as adaptable as possible; this includes a mobile version of the component to be ran on smartphone devices, as well as an augmented reality counterpart, meant to be accessed with smart glasses (e.g. Vuzix M400) in order to free the operator from additional devices and allow him to perform recycling procedures hands-free.

5 Case Studies

5.1 Reuse of PCBs and Recovery of CRMs

As part of the activities to support PCBs reuse and CRMs recovery in end-of-life operations, a collaboration with ERION enabled the application of the Circularity Toolbox in the mass electronics sector. To better understand how the toolbox addresses the current use-case, the end-to-end flow is described (see **Errore. L'origine riferimento non è stata trovata.**), together with an interaction diagram highlighting the synergies between the modules presented above.

The flow starts from BoL, where manufacturers can specify the criteria each PCB needs to meet in order to be considered recoverable. Such information is added into the Marketplace module. Next, the manufacturer may decide to share with trusted dismantling facilities information about the appliance disassembly operations. The recycling plant, on the other hand, can leverage the information shared by BoL actors to efficiently extract PCBs from appliance that matches the reusability criteria expressed from manufacturers. This procedure is carried out via augmented/virtual reality by means of the Advanced HMIs module to visualize the shared assets that are taken from the Dataspace. Once PCBs are available for reuse, the two parties can interact via the Marketplace module to arrive at an agreement, assisted by autonomous agent-based negotiation to support the decision-making process. The agreement is then signed, validated and finalized using state-of-the-art blockchain technology, ensuring transparency and traceability of the transactions, in order to be compliant with EU regulations in terms of Digital Product Passport (DPP).

The Frontend component of the Advanced HMIs module supports the recycling operator in assessing at a glance the classification outcome provided by the backend AI classifier (Backend component, presented next). The component is offered in various configurations to adapt to different scenarios and needs; this includes a mobile version of the component to be ran on mobile devices, as well as an augmented reality counterpart, meant to be accessed with smart glasses (e.g. Vuzix M400) in order to free the operator from additional devices and allow them to perform recycling procedures hands-free. The assessing feature described here is meant to seamlessly integrate with the functionalities offered in the ECU use-case, in order to link it with components disposal and provide all-around support for the recycling operators, regardless of the activity to carry out. In other words, this integration enables switching between reuse and recycling tasks with virtually zero effort, allowing dismantlers to include this workflow in their set of activities without the need to set up a dedicated auxiliary process.

The Backend component of the Advanced HMIs module (formerly known as the **AI PCB classifier**), has the objective of effectively sorting WEEE, more specifically printed circuit boards (PCBs) according to their value and provide access to the classification results to the rest of the toolbox (including the frontend component of the Advanced HMIs) in a standardized way. The core of this component lies in the aim of automatically assigning each PCB image to the class to which it belongs, selecting from pre-defined categories (classes) of interest, based on PCBs' value, thus improving sorting accuracy. To this extent image classification algorithms are implemented with the objective of minimizing the errors inherent in the manual sorting system that currently takes place in recycling plants, which relies on operator knowledge. This improves efficiency and ensures that PCBs are accurately sorted according to their respective classes. It also aims to maximize the correct reuse of materials, thereby promoting resource conservation and economic sustainability.

The analysis focuses on an initial proof of concept targeting Hard Disk drives, with the objective of expanding the initial dataset with all types of PCBs. For this first iteration, 457 pictures from the TRANSISTOR plant were gathered and four classes were identified together with operators, depending on the amount of precious chips and materials available on the boards: (Fig. 4). To achieve a usable balanced dataset, traditional techniques have been applied (data augmentation, resizing, sub-sampling) and the resulting dataset was then split 80% for training, 20% for validation.

The hyperparameter tuning process aimed to optimize the selected pre-trained model (ResNet CNN) for PCB image classification. Key hyperparameters adjusted

Fig. 4 A sample for each category of PCBs; highest value (top center), high value (bottom center), medium value (left), low value (right)

included batch size, epochs, number of unfrozen layers, optimizer type, and learning rate. Only the final layers were retrained to adapt to new categories while leveraging pre-trained knowledge. The tuning focused on improving gradient accuracy, balancing learning, and enhancing feature extraction. A search space was explored to minimize training and validation loss while ensuring smooth convergence. Dropout regularization with a 0.5 rate was applied to prevent overfitting. This combination of tuning and regularization enabled the CNN to generalize well to new data and accurately classify images across all PCB categories (Fig. 5).

In the final stage, the tuned model achieved a 65% accuracy on the validation set, with training halted after 20 epochs due to divergence in the validation loss, indicating potential overfitting. This relatively low accuracy suggests issues with data representativeness, as the training dataset may lack sufficient diversity to generalize effectively. Additionally, the limited dataset size might have restricted the model's ability to learn robust features. Overall, these results are considered satisfactory for the first iteration of the model, considering it is still a proof of concept to be further investigated in the second iteration with a higher amount of data available.

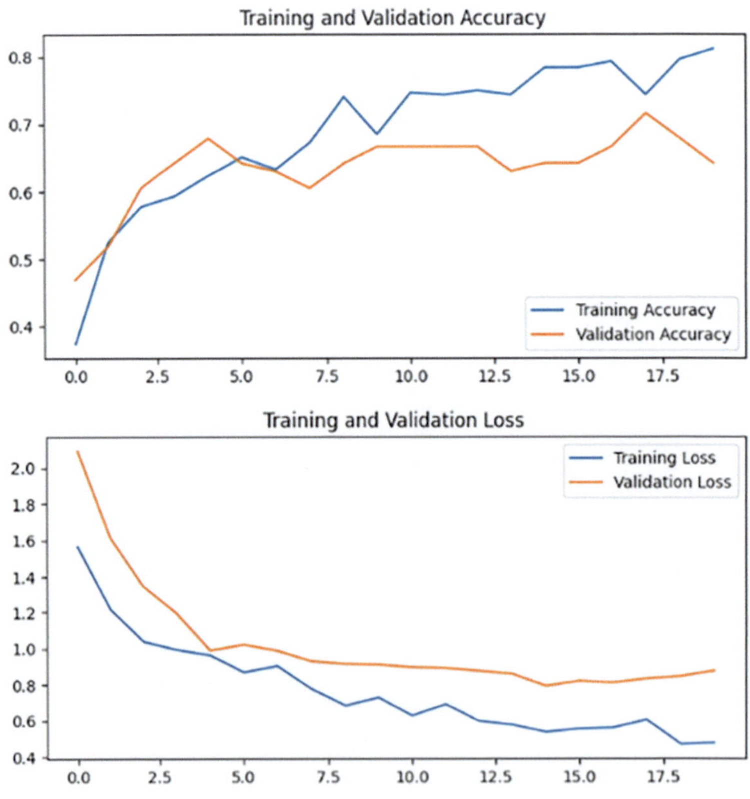

Fig. 5 Plot of accuracy and loss curves of the model on training and validation sets

5.2 Repair for ECUs

As part of the digital twin setup, a model for ECU (Electronic Control Unit) diagnostics in collaboration with BOSCH was developed. The whole use case of BOSCH is represented in the digital twin module of the toolbox as example for other use cases. In addition, it uses the LCS&CA Module of the toolbox for the environmental assessment.

The computer vision fault detection model is deployed locally to assess the binary status of seven critical ECU components, determining if they are functional or faulty. The goal of using machine learning here is to quickly identify non-repairable products, increasing sustainability and efficiency by avoiding unnecessary resource use for transportation, cleaning, and sorting. In essence, it asks: "Is this product worth repairing?" This approach could replace time-consuming manual inspections and significantly save resources.

The project has two well-known yet challenging limitations: class imbalance and training data scarcity. The class imbalance refers to the fact that non-faulty instances greatly outnumber faulty ones, and the different classes of faults are not equally distributed. The number of samples (ECUs) provided was critically small. To address class imbalance, we experiment with two approaches:

(i) Data-level approaches, including Random Under-Sampling, Random Over-Sampling, Dynamic Sampling, BGAN [6], and augmenting specific areas of an image with elastic distortions or noise injection (via region of interest (ROIs) [7], and (ii) Algorithm-level methods, such as customized loss functions, CSDNN, output thresholding, and category entries [8].

The issue of data scarcity is also being tackled through two approaches: (i) Dataset augmentation by acquiring new samples through a combination of existing ones via ROI cropping and creating new samples (ECUs) by physically altering the existing ones, and (ii) Transfer learning.

The motivation behind the image capturing and labelling algorithm created during the initial stage stems primarily from the need for fail-safe image files, which are 8-character binary codes with a unique ECU identifier as a suffix, and the desire for homogeneity in image capturing settings (e.g., exposure time). The main challenge in image capturing was the reflective nature of some ECU components, which caused conflicts with non-reflective parts when adjusting exposure times (see Fig. 6 left). To address this, we devised multiple approaches, such as combining two images with different exposure times by cropping and using HDR techniques with some contrast enhancement (see Fig. 6 right). As a result, multiple datasets are created and compared using model accuracy metrics.

By identifying the "repairable" parts through computer vision, the next step is to relate the parts to the Digital Twin and GRETA Module to assess the environmental performance. This enables linking the results of the sustainability assessment to the real world application.

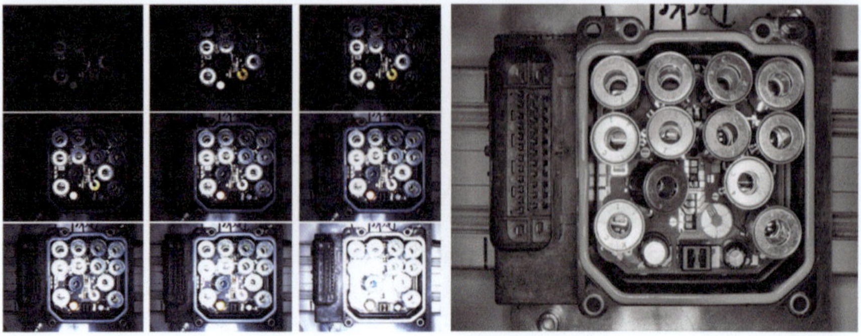

Fig. 6 Left: 9 photos with different exposure times. Right: HDR derived image from those 9 photos

6 Outlook

Following the reuse and repair phases of its lifecycle, a component will eventually need to be discarded and recycled.

Further detail on the link between product design considerations, product compositional data requirements and linking disassembly, repair and recycling processing, to assess and optimize EoL recovery from these products from an environmental and sustainability point of view, can be found in the chapter on Sustainability and Circularity Assessment & Advisory Methodologies and Services. In this Chapter, rigorous technology—and physics-based methodologies for the assessment of recycling, calculation of recycling KPIs, optimization of recycling system flowsheets and Design for Recycling and disassembly advisory on this rigorous basis will be discussed and demonstrated (see also [9, 10]).

Ultimately, digital tools have the potential to enhance sustainability along the life cycle by addressing different difficulties and knowledge gaps. The circularity toolbox contains different tools that can be used throughout the life cycle of physical products, showcasing the unique needs that each stage has. Together with the analysis presented in the chapter on Sustainability and Circularity Assessment & Advisory Methodologies and Services nearly all sustainability challenges of this time can be tackled. The tools are validated with the pilots, as described in the respective chapters, i.e. the Ecodesign of ECUs, Ecodesign of TMPS, Ecodesign of IME and Obsolete PCB sorting.

References

1. Manhart A, Vogt R, Priester M et al (2019) The environmental criticality of primary raw materials—a new methodology to assess global environmental hazard potentials of minerals and metals from mining. Miner Econ 32:91–107. https://doi.org/10.1007/s13563-018-0160-0

2. Penaherrera F (2024) Analysis of interactions between raw material and energy demands for data centers. Dissertation, Carl von Ossietzky Universität Oldenburg
3. Nasr N (ed) (2024) Technology innovation for the circular economy. Recycling, remanufacturing, design, system analysis and logistics. Wiley, Scrivener Publishing, Hoboken, NJ, Beverly, MA
4. European Parliament (2023) Proposal for a REGULATION OF THE EUROPEAN PARLIAMENT AND OF THE COUNCIL establishing a framework for ensuring a secure and sustainable supply of critical raw materials and amending Regulations (EU) 168/2013, (EU) 2018/858, 2018/1724 and (EU) 2019/1020
5. Perossa D, Rosa P, Terzi S (2024) A systematic literature review of existing methods and tools for the criticality assessment of raw materials: a focus on the relations between the concepts of criticality and environmental sustainability. Resources 13:131. https://doi.org/10.3390/resources13090131
6. Asokan V (2021) Handling class imbalance using generative adversarial network (GAN) and convolutional neural network (CNN). University of Reading, Masterthesis
7. Simard PY, Steinkraus D, Platt JC (2003) Best practices for convolutional neural networks applied to visual document analysis. In: Proceedings/seventh international conference on document analysis and recognition, 3–6 Aug, 2003, Edinburgh, Scotland. IEEE Computer Society, Los Alamitos, Calif., pp 958–963
8. Wang H, Cui Z, Chen Y et al (2018) Predicting hospital readmission via cost-sensitive deep learning. IEEE/ACM Trans Comput Biol Bioinform 15:1968–1978. https://doi.org/10.1109/TCBB.2018.2827029
9. Meskers C, Worrell E, Reuter MA (eds) (2024) Handbook of recycling. State-of-the-art for practitioners, analysts, and scientists, 2nd edn. Elsevier, Amsterdam, Oxford, Cambridge, MA
10. van Schaik A, Reuter MA (2024) Simulation-based design for recycling of car electronic modules as a function of disassembly strategies. Sustainability 16:9048. https://doi.org/10.3390/su16209048

Sustainability and Circularity Assessment and Advisory Methodologies and Services

Alessandro Fontana, Siro Dell'Ambrogio, Veronica Dosi, Simone Fasola, Giuseppe Landolfi, Christian Trisolini, Antoinette van Schaik, Markus A. Reuter, Lisa Dawel, and Alexandra Pehlken

Abstract To guide a decision-making process oriented to design and produce more sustainable and circular products and components, a comprehensive sustainability assessment and advisory framework was developed and customized for each pilot project within the CIRC-UITS project. To ensure it addressed the specific needs of each case (ECU (re)design, design of tire sensors, flexible electronics, and PCB

A. Fontana (✉) · S. Dell'Ambrogio · V. Dosi · S. Fasola · G. Landolfi · C. Trisolini
SUPSI, Institute of Systems and Technologies for Sustainable Production (ISTePS), Lugano, Switzerland
e-mail: alessandro.fontana@supsi.ch

S. Dell'Ambrogio
e-mail: siro.dellambrogio@supsi.ch

V. Dosi
e-mail: veronica.dosi@supsi.ch

S. Fasola
e-mail: simone.fasola@supsi.ch

G. Landolfi
e-mail: giuseppe.landolfi@supsi.ch

C. Trisolini
e-mail: christian.trisolini@supsi.ch

A. van Schaik · M. A. Reuter
MARAS—Material Recycling and Sustainability B.V., The Hague, The Netherlands
e-mail: a.vanschaik@maras-bv.nl

M. A. Reuter
e-mail: markusandreas.reuter@sms-group.com; markusandreasreuter@icloud.com

L. Dawel · A. Pehlken
FuE Bereich Produktion, R&D Division Manufacturing, OFFIS e.V.—Institut für Informatik, Oldenburg, Germany
e-mail: lisa.dawel@offis.de

A. Pehlken
e-mail: alexandra.pehlken@offis.de

P. Rosa and S. Terzi (eds.), *Circular Approaches for the Ecodesign, Repair and Remanufacturing of Car and Mass Electronics*,
PoliMI SpringerBriefs, https://doi.org/10.1007/978-3-032-04301-6_4

recycling and repair), targeted activities were organized with the industry partners, including workshops aimed at understanding sustainability and circularity requirements. These collaborative sessions underscored the necessity for rigorous and advanced methodologies, such as an enhanced Material Circularity Indicator (MCI), a repairability and repair assessment, and AI-supported interpretation of LCA results and recycling assessments. These methodologies were particularly valuable for hotspot identification, as recommended by ISO 14044 [1], and for true Design assessments and Design for Recycling, based on industrial End-of-Life processing of complex multi-material electronics. Drawing from these insights, the most appropriate methodologies and indicators were selected to meet the unique needs of each pilot. Alongside the framework development, critical moments were identified to provide advisory support, specifically aimed at enhancing the Eco-Design process, LCA, and recycling efforts. This support was integral to informed decision-making in sustainability evaluations. The implementation phase then focused on deploying the tools necessary to operationalize this framework. This included high-level descriptions of tools such as GRETA, recycling simulations, and newly developed AI-driven assessment and advisory components.

Keywords Life Cycle Assessment (LCA) · Life Cycle Sustainability Assessment (LCSA) · Circular Economy · Recycling · Simulation modelling · Repairability · AI-Powered Decision Support

1 Overview of the Life Cycle Sustainability and Circularity Assessment (LCS&CA) Framework

This section aims to present the most suitable assessment methodologies to support decision-making in the context of circular electronic products. For the sustainability dimension, three methodologies are proposed to address environmental, economic, and social aspects. Regarding the recycling dimension, specific methodologies are introduced to assess the efficiency and impact of recycling processes. Finally, for the circularity dimension, the evaluation focuses on material circularity and product repairability.

1.1 Sustainability Assessment Framework: Environmental, Economic, and Social Dimensions

The CIRC-UITS project employs three key methodologies for assessing environmental, economic, and social impacts: Life Cycle Assessment (LCA), Life Cycle Costing (LCC), and Social Life Cycle Assessment (SLCA). Each of these approaches contributes to the measurement of the impacts of circularity in the semiconductor and electronics sectors by assessing the different stages of the life cycle.

Life Cycle Assessment (LCA)—LCA is used as the primary methodology for assessing the environmental impacts of product systems, following [1]. It considers a product's entire life cycle, from raw material extraction through production, use, and final disposal, including processes such as recycling and reuse. In the CIRC-UITS project, LCA focuses on a "cradle-to-cradle" system boundary, particularly emphasizing the Begin-of-Life (BoL) and End-of-Life (EoL) phases. These stages are critical for advancing circularity, as they evaluate the potential for reuse, repair, and recycling of semiconductors, while minimizing environmental impacts. In the CIRC-UITS project, the Product Environmental Footprint (PEF) [2] methodology has been selected for conducting the LCA. Although based on the principles of LCA, PEF offers more specific guidelines and criteria, allowing for greater comparability of results between similar products.

Life Cycle Costing (LCC)—LCC is applied to assess the economic costs associated with a product's entire life cycle, following ISO 15686-5 guidelines [3]. CIRC-UITS adopts the conventional LCC (cLCC) methodology that excludes externalities such as environmental or societal costs. The LCC approach in CIRC-UITS includes various cost items like personnel, materials, energy, and End-of-Use (EoU) costs such as disassembly and waste management. This methodology helps quantify the financial sustainability of circularity strategies, ensuring that investments are both viable and profitable across the product life cycle.

Social Life Cycle Assessment (SLCA)—SLCA in CIRC-UITS is based on the UNEP framework, [4] incorporating insights from previous projects such as Circ-Thread [5] and TREASURE [6]. It evaluates the social impacts of product systems across their life cycles, focusing on stakeholder categories like workers, local communities, and society as a whole. The SLCA methodology integrates with the LCA framework, allowing for the evaluation of both environmental and social impacts simultaneously. The PSILCA and SOCA databases [7],[1] are employed to quantify social risks and opportunities, offering detailed assessments that help identify areas for social improvement along the supply chain.

Together, these methodologies provide a comprehensive assessment framework that balances environmental, economic, and social dimensions.

1.2 Recycling Assessment

A rigorous and objective approach is required to protect and safeguard the recycling and metallurgical processing industry at the heart of the Circular Economy (CE),

[1] "SOCA reference": GreenDelta (2024) SOCA v3.0—Social add-on to ecoinvent (based on ecoinvent v3.10 and PSILCA v3.1.1). openLCA Nexus. Available at: https://nexus.openlca.org/database/soca. Accessed 4 September 2025.

whatever the CE business model is Reuter et al. [28]. Gleaning from the knowledge of minerals engineering and process metallurgy, simulation and process control tools, and digitalization platforms, this section will show that the key to enabling an understanding of the CE is a product(mineral)-centric description as well as a thermochemical and therefore exergetic understanding and characterization of all complex associated and functionally linked particles, modules, materials, and so forth in the various stages of their respective CE journey.

The predictive nature of simulation models allows for the physics-based estimation of how life-cycle systems respond to changes in, among others, feed material compositions, process configurations, operating conditions, and the technologies used. Process simulation thereby enables Design for Recycling (DfR) [17, 18]. The system-wide effects of product design changes on resource consumption and sustainability can be evaluated as early as during the product design phase to maximize recyclability. Furthermore, where possible, the use of process simulation models allows for residues formed in production and recycling processes to be engineered to meet secondary resource specifications, in the process maximizing resource efficiency.

The material and energy consumption, emissions, and residue formation data generated through process simulation for complete supply chains can be transferred into various other tools to conduct further analyses. Among these are methods used to assess environmental, economic, and social impacts. Environmental LCAs, for example, can in some cases be overly reliant on environmental databases in which some datasets are significantly outdated and others do not exist yet because of rapid technology and process development, especially with the increasing focus on more targeted, high-quality recycling. Process simulation can enhance the quality of such assessments by providing up-to-date inventory data [24]. The flexibility to generate new datasets for alternative scenarios quickly can contribute to identifying process and technology options that would maximize the recovery of potential secondary resources at the highest possible purities within the physics-based, environmental, economic, and societal constraints the system is bound by.

The indicators obtained are listed below and can be used to effectively communicate reliable process and sustainability information to industry, consumers, and policymakers to support decision-making and drive progress toward sustainability and circular economy.

Indicators/KPIs calculated and derived from simulation models

- Recycling Index (RI) (see Fig. 1): Total recycling rate of a product or part based on all materials/compounds present (mass or %) (based on full mass balances, hence can also be used to provide an indicator/value for 'waste'/losses/emissions from recycling)
- Material Recycling Index (MRI) (see Fig. 1): Individual material recycling rate for all materials/elements present in a product/part (mass or %) (based on full mass balances, hence can also be used to provide an indicator/value for 'waste'/losses/emissions from recycling) timescaled Material Quality Indicator (MQI): Quality/purity of recovered materials—CE level of application of recycling products

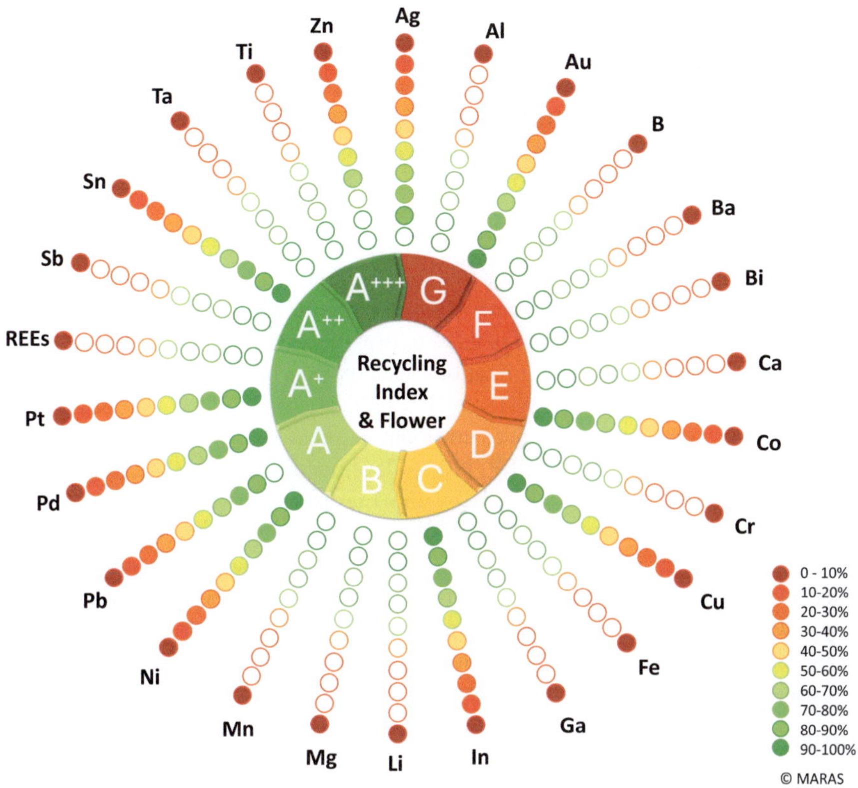

Fig. 1 Recycling index and material recycling flower visualizing the total and material recycling rate [18]

- Energy/exergy Circularity Indicators (ECI): Energy usage and recovery in recycling and exergy quantifying the CE performance of the recycling system on a physics (2nd law of thermodynamics) basis

1.3 Circularity Assessment Framework

Overview of Existing Circularity Assessment Methods

In the CIRC-UITS project, several methodologies have been evaluated to assess the overall circularity of products and materials. These methodologies go beyond just the environmental or economic impacts, offering a more comprehensive view of circularity by focusing on resource efficiency and material flow. Three key methodologies have been selected and introduced in the following to measure circularity as a whole.

Material Circularity Indicator (MCI)—Developed by the Ellen MacArthur Foundation, [8] the MCI evaluates the degree to which a product's materials are part of a circular system. It measures how much a product minimizes the use of virgin materials and reduces waste through recycling and reuse. The MCI assigns a value between 0 and 1, where 0 represents a purely linear process with high material loss, and 1 reflects a fully circular product with no virgin material and full recycling. MCI helps quantify the linear flow versus restorative flow of materials within a product's lifecycle and includes consideration of factors like the product's lifespan and intensity of use.

Product Circularity Indicator (PCI) [9]—The PCI builds upon the MCI by focusing on the circularity at a component or product level, taking into account factors like reused content, the recovery of materials within the system, and prioritizing reuse over recycling. While the PCI is intended to provide a more granular view than the MCI, it also addresses specific limitations such as the circularity of material flows and their integration within a system. This indicator particularly emphasizes the importance of component reuse over raw material recycling.

Circular Footprint Formula (CFF) [10]—Developed as part of the Product Environmental Footprint (PEF) initiative by the European Commission, the CFF is designed to account for material circularity while also incorporating energy use and disposal impacts at the end of life. The CFF evaluates the benefits and burdens associated with recycling processes, energy recovery, and the use of secondary materials, offering a more comprehensive understanding of the circular performance of products. It considers both material and energy contributions, making it a holistic approach to evaluating circularity in complex product systems like semiconductors.

Advancement of Existing Circularity: Repairability, Repair and Exergetic (Recycling) Assessment

To improve existing CE methodologies such as the MCI, PCI, and CFF methodologies, the CIRC-UITS project also incorporates specific circularity approaches, such as the Repairability Assessment, the newly defined Repair Assessment and Recycling Assessment methodologies, which provide rigorous EoL assessment as well as a physics-based approach to circularity on the basis of exergy.

As the existing Circularity Indicators as listed above do not find their basis in physics and rely on generic approaches and simplified non-product related databases, which do not recognise the role of multi-material design and physics and thermodynamics in recycling, these indicators do not have the capability to correctly quantify Circularity and lack sufficient detail to truly discern between different design, recycling and repair for Circularity options. In Sect. 3 it will be elaborated on how the process simulation and underlying data and KPIs provides improved and design and recycling process specific input to existing MCI (and others) c and provides additional values to improve the existing CE indicators with a rigorous basis. Within this project, both improvement of existing indicators (PCI, MCI etc.) will be explored, as well as the introduction of a new exergy-based indicator, expanding and improving the basis for Circularity Indicators on a more extended and true system physics and 2nd law of thermodynamics-based approach to current indicators.

The repairability and repair assessment methods do not evaluate circularity in its entirety but rather focus on specific aspects, such as the product's ability to be repaired.

The proposed repairability assessment follows the EN 45554 standard, [11] which provides generic methods to evaluate a product's ability to be repaired. The process focuses on returning faulty products to functional conditions and involves several stages. The EN 45554 standard outlines a structured approach that begins with identifying priority parts based on their likelihood of failure and importance to the overall functionality of the product. Once the critical components are identified, various criteria, such as disassembly depth, fastener types, and availability of spare parts, are evaluated. These criteria help assess the ease with which repairs can be performed and spare parts replaced, which directly impacts a product's overall repairability score. To quantify repairability, the standard provides a methodology for assigning numeric values to each criterion, which are then aggregated into a single repairability score. Higher scores indicate better repairability, making it easier for consumers and manufacturers to prioritize repairs over replacement. This methodology promotes product longevity and reduces waste, as it encourages design choices that facilitate repair rather than disposal.

Repair assessment is not covered by the EN 45554 repairability standard. In this work, a repair assessment approach is defined by MARAS and SUPSI in which the material, energy, environmental and social effects of repair actions can be quantified by the application and combination of the LCS&CA methodologies as discussed here and applied in the CIRC-UITs project. This approach will be discussed in Sect. 3.3.

Circularity Assessment Framework Conclusion

Although these CE methodologies provide CIRC-UITS with robust tools to assess circularity, they have been deemed insufficient to meet some specific needs of the project pilots, as illustrated in Sect. 2.1. Consequently, in Sect. 3 the modified methodologies are described, which have been adapted to better address the requirements of the individual use cases and which will bring improvement to the field of Circularity and Repair Assessment by the application of the developed LCS&CA methodologies.

2 LCS&CA and Advisory in Project Pilots

The LCS&CA framework was customized to meet the unique requirements of the project pilot by addressing key needs and establishing a coherent methodology. This customization process aimed to align sustainability and circularity indicators with the pilot's specific goals, focusing on the integration of an advisory function. The objective was to create a tailored framework capable of supporting non-experts in evaluating and improving sustainability and circularity impacts through informed decision-making, facilitated by AI-driven advisory components.

2.1 Methods to Obtain a Customized LCS&CA Framework with Advisory Function

The customization approach for the LCS&CA framework was organized around four key pillars: the development of a comprehensive inventory table, the identification of assessment and advisory needs through dedicated workshops and stakeholder engagements, the creation of detailed user stories to illustrate interactions with the advisory functions, and the careful selection and prioritization of relevant indicators. This structured approach ensured to build a cohesive and effective LCS&CA framework, tailored to the unique requirements of each project pilot. To illustrate the customization approach, an example based on one of the four pilots is presented for each key pillar listed before. The selected pilot focuses on improving the sustainability and circularity of Electronic Control Units (ECUs), serving as a representative case for the LCS&CA framework's application.

Development of Inventory Table

The inventory table of each pilot served as the foundation for understanding the activities, goals, and challenges faced by each pilot. The inventory table was designed to capture key details about the life cycle phases of each product, focusing on areas such as design, manufacturing, and end-of-life (EoL) processes. The inventory table preparation involved a close collaboration with stakeholders to ensure that the collected information was accurate and comprehensive.

The aim of the inventory table was to outline all relevant activities and processes, helping the project team identify which stages of the product lifecycle had the most significant impact on sustainability and circularity. Additionally, it provided a basis for understanding the operational context of each pilot, which was crucial for customizing advisory functions.

Concerning the illustrative pilot, the inventory table provided a detailed breakdown of the ECU's lifecycle stages, covering key areas such as design, manufacturing, use, and end-of-life (EoL) processes. The table captured crucial information about activities and material flows within each lifecycle stage. For instance, in the design phase, activities focus on selecting sustainable materials and optimizing designs for disassembly. In the manufacturing phase, the emphasis is on reducing energy consumption and material waste through process improvements.

Collection of Assessment and Advisory Needs

Following the development of the inventory table, the next step involved identifying the specific needs of each pilot, focusing on both assessment and advisory requirements. Needs identification was conducted through workshops and stakeholder consultations, gathering insights into key decision-making points and sustainability objectives.

The needs were divided into two main categories: assessment needs and advisory needs. Assessment needs referred to the requirements for measuring and evaluating the sustainability and circularity performance of each pilot. These needs included data collection requirements, measurement techniques, and specific assessment methodologies. Advisory needs, on the other hand, involved the support that pilots required to make informed decisions based on sustainability indicators. This included real-time decision support, trade-off analysis, and recommendations for design improvements.

Starting from the inventory table, specific assessment and advisory needs have been identified for the ECU case. The assessment needs primarily revolved around evaluating the environmental impacts of various materials and calculating lifecycle costs. Advisory needs, on the other hand, are focused on providing real-time feedback on design decisions, especially regarding the trade-offs between recyclability and material durability. One of the key advisory requirements identified was the need for support in comparing material options for soldering. The pilot required insights on the potential environmental impact of each material choice, as well as guidance on improving the repairability and recyclability of the final product.

Development of User Stories

Once the needs were identified, the focus is shifted to the development of the user stories that are meant to define how stakeholders would interact with the LCS&CA framework and its advisory functions. User stories are short narratives that describe specific scenarios where stakeholders require advisory support. They are intended to illustrate how the advisory function would be used in practice to address key decision-making challenges.

The development of user stories involved closely engaging with stakeholders to understand their workflows and decision-making processes. Each user story captured a specific interaction, such as an engineer needing to select a material with high recyclability or a project manager evaluating the environmental impact of a new design concept. These user stories were instrumental in shaping the advisory functions, ensuring that they provided relevant and actionable insights in real-world scenarios.

Figure 2 represents the user story for the assessment and advisory process of the illustrative pilot. Initially, inventory data from the lifecycle stages of the current ECU is collected, such as: BOM-related data, manufacturing process data, energy consumption data for desoldering, and data required for the recycling assessment. Using this data, various assessments are conducted, specifically environmental, circularity, and End-of-Life assessments. The assessment provides impact results, introducing the first advisory functionality: the tool identifies the system's Hot Spots, supporting designer in prioritizing eco-design actions. The second advisory functionalities that come into play here are the chatbot and best customization features. Both are AI-powered and designed to support the designer's decision-making process, assisting in the development of more sustainable designs. Once the new product design is conceived, a comparison is made between the old and new designs, and their impacts are evaluated. If the sustainability targets are met, the new design is approved.

Indicators Selection

The final step in the LCS&CA customization process involved selecting relevant indicators to measure sustainability and circularity performance. Various selection

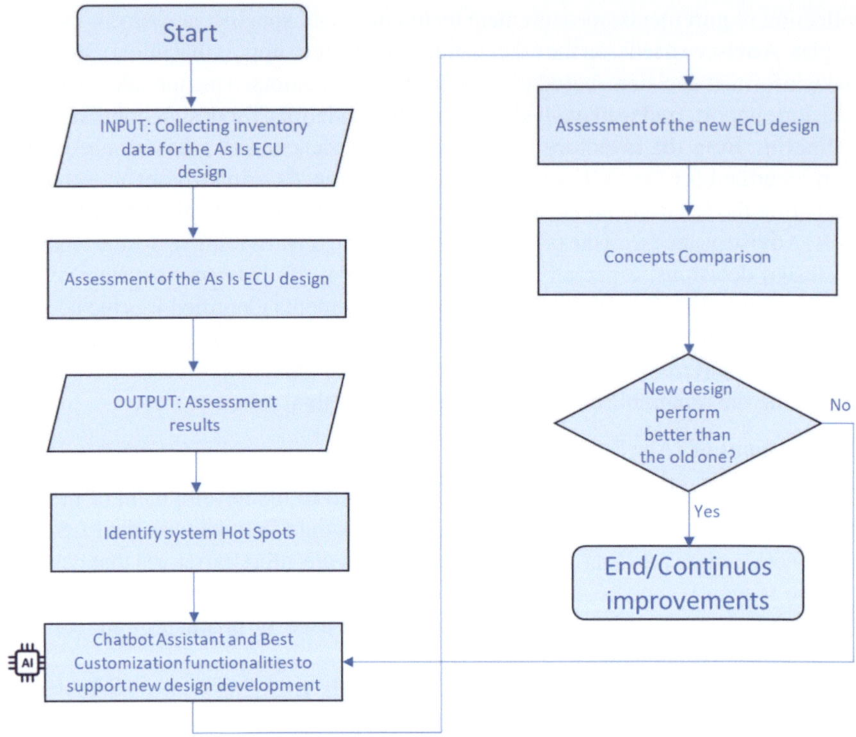

Fig. 2 LCS&CA and advisory user story—Pilot 1

methods were employed among stakeholders to prioritize indicators based on their alignment with the project's sustainability goals. Environmental, economic, financial, social, and circularity indicators were all considered to ensure a comprehensive assessment of each pilot's performance.

The results of the surveys were analysed to identify high-priority indicators aligned with the specific goals of the pilots. Environmental, economic, and social indicators were selected through a structured methodology involving surveys and the Analytical Hierarchy Process (AHP). Additionally, a workshop emphasized the need for circularity assessment, leading to the inclusion of a modified MCI/PCI, a repairability score, and recycling-related indicators.

This selection process ensured that the chosen indicators accurately reflected the priorities of each pilot and provided meaningful insights into their sustainability and circularity performance.

In the ECU pilot, selecting environmental indicators was essential for assessing the sustainability of innovations in ECU components. The process focused on aligning with project goals and addressing the pilot's specific needs. The top-rated indicators for the illustrative pilot are: Global Warming Potential (kg CO_2 eq.); Water Use (m^3 water eq. of deprived water); and Minerals and Metals Resource Use (Abiotic resource depletion, kg Sb eq.). In agreement with stakeholders involved, it was decided not to include economic, financial, and social indicators in the assessments of this pilot. Instead, a set of indicators was selected to measure circularity, including an enhanced version of the MCI, repairability indicators, and recyclability indicators.

2.2 *Key Workshop Outcomes and Insight*

Needs for New Repair and Repairability Assessment Methods

During the CIRC-UITS project workshops, repair assessment and repairability assessment emerged as a key concern for circularity assessment. All pilot participants identified it as a critical factor, highlighting the need for a more comprehensive evaluation methodology. While existing indicators like the Material Circularity Indicator (MCI) and Product Circularity Indicator (PCI) are adequate for most circularity assessments, they have gaps in covering repairability, reuse, and remanufacturing. Furthermore, these indicators are not well-suited to assess the recycling impacts with a physics-based approach, such as exergy considerations, which provide a more rigorous evaluation of End-of-Life (EoL) processes and deeper insights into circularity.

Another finding was the shift from measuring repairability as a standalone metric to considering the broader impacts of repair within the product life cycle. As a result, the project differentiated between two types of assessments:

- Repairability Assessment: Focuses on the product's inherent ability to be repaired, providing metrics on how easily and effectively it can undergo repair.
- Repair Assessment: Takes a more comprehensive approach, analysing environmental, social, economic, and recycling impacts associated with repair scenarios. It includes LCS&CA of scenarios where the repair phase is integrated into the product life cycle, evaluating impacts such as spare parts production, dismantling processes, repair operations, and associated resource use and emissions.

To address these gaps, the project explores three different approaches: (i) Improved version of the PCI/MCI indicators, as discussed in Sect. 3.1; (ii) Repair assessment based on LCS&CA methodologies, as discussed in Sect. 3.3; (iii) Repairability assessment based on new standards, as discussed in Sect. 3.4.

The Role of AI in LCS&CA, Advisory and Repairability Assessments

AI emerged as a pivotal solution for addressing complexities in repairability assessments. The integration of AI can streamline data analysis and enhance decision-making processes, such as prioritizing the reparability of critical components based on failure frequency, available repair information, and other criteria. The CIRC-UITS project has also explored AI-driven advisory tools like chatbots, which can assist stakeholders in understanding and interpreting results from Life Cycle Assessments (LCAs). These chatbots can compare multiple scenarios, guide users through assessment outcomes, and suggest sustainable options based on key findings.

Moreover, AI techniques such as machine learning and deep learning were identified as instrumental in refining repairability scoring systems. For example, by analysing historical data on product failures and repair outcomes, AI can help recalibrate repairability criteria or weightings to ensure alignment with real-world conditions. This approach is expected to improve the robustness and practicality of the methodologies adopted.

3 Emerging Circular Methodologies

3.1 Improved Version of the MCI Indicator

As the existing Circular Methodologies as listed in Sect. 2 of this Chapter do not rely on product and process specific data and are therefore too generalistic and incorrect, these methodologies have been improved and advanced by bringing in the rigour of the defined LCS&CA methods of the CIRC-UITS project. In this sense, the project highly contributes to the improvement of Circular Assessment methods and the rigour, reliability and technological and physics-based soundness of the underlying indicators. Repair within the MCI/PCI indicators should be addressed by accounting for the additional material and energy flows involved in repair processes. The enhanced version of MCI/PCI should integrate the rigor of LCS&CA methodologies, further strengthening the robustness of these indicators.

Indicators and data as derived from the LCS&CA tools provide input to MCI (and others). The recycling simulation tool will provide improved and design and recycling route specific data to the MCI indicator on recycled material/recovered EoL materials, waste from recycling, energy required for recycling, etc., while the LCA tool will contribute data on energy and waste from primary production, and the content of virgin materials, thereby refining the overall assessment process and ensuring alignment with broader circularity goals while preserving the integrity of existing frameworks. This is listed in Table 1.

3.2 Exergy as CE Indicator

The aim of a circularity approach and circularity indicators should be to identify and minimize residues and losses, i.e., to minimize the creation of entropy, across whole value chains of the CE in addition to closing material loops through EoL recycling. The present digitalization platforms have evolved significantly to estimate the bulk, minor elements, technology elements, metal, alloys, and material flows in addition to

Table 1 OpenLCA MCI parameters and improvement based on LCS&CA methods

MCI parameters in OpenLCA	Unit	LCS&CA methodology for improvement of MCI
Energy required for primary production	MJ	LCA
Energy required for recycled production	MJ	Recycling simulation tool
Recovered EoL material	kg	Recycling simulation tool
Recycled material	kg	Recycling simulation tool
Total waste produced (W)	kg	LCA + Recycling simulation tool
Virgin material (V)	kg	LCA + Recycling simulation tool (for EoL virgin material need)
Waste from recycling processes (Wc)	kg	Recycling simulation tool
Waste from the production of feedstock, for second life (Wf)	kg	LCA

the exergy and energy flows of the complete CE system with its various stakeholders (Reuter 2016). These stakeholders in the product chain can been linked in a uniquely digitalized platform to estimate the exergy destruction of a significant part of the CE system [24].

Exergy is a key aspect that defines the efficiency of the CE [20, 21, 22] and was unfortunately not discussed by Reck and Graedel [26], leading the CE discussion down an incorrect path. The inescapable second law of thermodynamics (2LT) governs the efficiency (or inefficiency) of every system, i.e. material, energy losses, specifically the entropy. In essence, the 2Lt suggests that 100% RE is a physical impossibility with respect to both energy (1LT and 2LT) and matter (through dilution, dissipation, imperfect separation etc.): irreversible losses are an inevitable part of industrial CE systems. Thus, losses and residues of uneconomic value are major causes of open CE loops, and closed loop recycling, i.e., the recycling of the same quality of material back into the same product it originated from, is the ideal. Generally, the lack of product-centric thinking and the lack of exergy and metallurgical detail in the CE discussions render these deliberations of CE rather unhelpful in understanding and providing quantified detail and innovative solutions to minimize the true losses from and hence to maximize the RE of the CE system and quantify economically CE business models.

The simulation-based approach underlying the recycling system assessment as discussed in this section provides the solution missing in the usual analysis of the CE by providing physics- based recycling standards based on exergy (kW) and energy (kW), which can also be derived from the simulation models in addition to the various other recycling KPIs such as Total and Material Recycling Indices, which are expressed in kg/h, t/h or % and Material Quality Indicator (MQI). This allows for a dynamic (flexible over time) and reliable harmonization method to calculate recycling rates and the resource efficiency of the CE system, in which the influence of EoL processing and product type/design are considered, based on the simultaneous calculation of mass, energy, and exergy balances for the processing of all materials and elements, compounds, alloys, etc., present in products. It permits the calculation of material-specific recycling rates, quality, energy and exergy depending on, e.g., product, product category, and design in terms of complex natural geological and complex designed mineral/functional material mixtures within the multi-element context of techno-economically viable flowsheets as discussed in detail in the Handbook of Recycling [25]. This rigorous and flexible simulation-based approach, in which design parameters are included, is based on industrial process physics, mass and heat transfer processes, reaction kinetics, and thermodynamics, as represented by, e.g., the software packages HSC Sim [23] and FactSage (www.factsage.com).

This approach to quantify CE is already available, as also demonstrated in recent work. CE indicators based on exergy will be applied within the CIRC-UITS project to quantify and optimize CE for different pilots. Design for Recycling and Design for Modularity advisory, recommendations for disassembly and optimal recycling flowsheet selection and configuration will be defined by assessing current and innovative designs within the project. The ever-present 2LT dictates that continuous reuse, reprocessing, renewal, and loss-less recycling cannot be achieved due to the degradation of the quality of energy (and thus materials). Losses and residues can be reduced but cannot be eliminated. Including this principle will stimulate and support impactful innovation on a true industry basis and will link the different stakeholders at BoL and EoL.

3.3 LCS&CA of the Repair

The EN repairability standard focusses on the assessment of a product in terms of ability of being repaired based on soft/estimated rankings. CE and a repair focused approach demands that the consequences and benefits of repair actions and options can also be assessed on an objective and rigorous basis, which provides insight into material savings and demands, recycling, life-time extension, environmental impacts and savings etc. of repair and recycling of products and parts. The hard quantified values/KPIs as derived from the defined and advanced LCS&CA methodologies within this project provide the input for such repair assessment as has been proposed by MARAS in the CIRC-UITS project and which brings together LCA & Recycling tools as discussed in this chapter to do so. This section will elaborate on the repair assessment and provide examples on this. The following list of Repair Assessment KPIs and factors have been defined by MARAS in order to be applied in the CIRC-UITS project for some of the pilots to support CE assessment from a repair perspective, linked to the LCS&CA indicators for the other life-cycle stages (see Table 2 and Fig. 3).

By using a time-scale for normalization of the material/resource and energy/exergy demands and recoveries for non-repairable versus repairable products/parts, which could have different life-times, these design options and their impact can be compared relative to a similar time-scale (e.g. impact per year of use or comparable).

Table 2 Repair indicators supported by recycling simulation tool and LCS&CA assessment

KPI name	Unit	Methodology
Saved (primary) resources	kg/kW	Recycling simulation tool (+ pilot data)
Recycling rates/energy balances/recovered material quality of recycling for comparison with repair	Kg/RR%/kW	Recycling simulation tool (+ pilot data)
Creation of remainders/ 'residues' and recyclability thereof	kg, RR%, kW	Recycling simulation tool
Disassembly time/costs	sec/€	Pilot data
Energy consumption of disassembly and repair	MJ/kW	Data from pilots
Expected life-time extension as a result of repair	Years	Data from pilots
Effects of human intervention in disassembly	To be determined	Social impact analyses, data from pilots
Material need/input for repair and waste created during repair (e.g. use of solvents etc.)	kg/kW	Data from pilots
Exergetic assessment recycling and repair (with respect to EoL)	kW	Recycling Simulation Tool
Environmental impact of materials needed for repair	According to PEF's indicators	LCA/pilot data
Environmental savings of extended life-time and saved resources due to repair	According to PEF's indicators	LCA
Time factor (to capture extended life time of materials in repaired part)	KPIs (impact)/year	Pilot data/empirical data

In Fig. 3 the role of time scale and LCS&CA KPIs/ factors in repair assessment are presented.

3.4 *Repairability Assessment*

The project explored the EN 45554 standard, which provides a structured methodology for evaluating product repairability. The standard outlines steps to identify priority parts and assess repair-related criteria, such as disassembly depth, spare parts availability, and required tools. However, pilot feedback emphasized the need for a

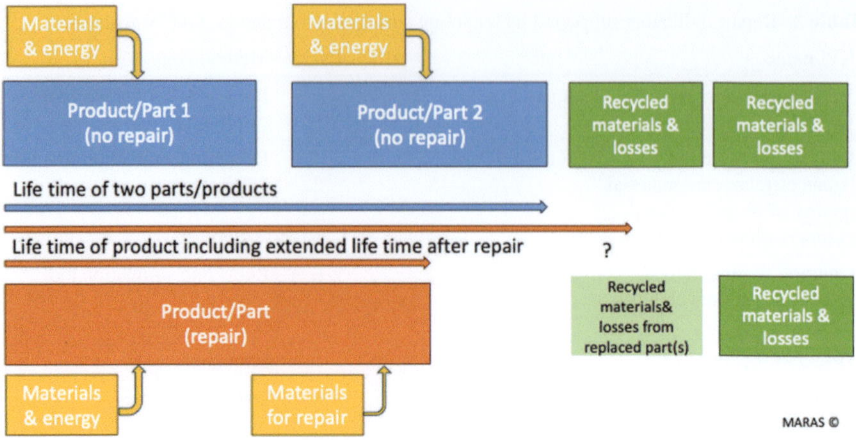

Fig. 3 The role of time scale and LCS&CA KPIs/factors in repair assessment (MARAS)

methodology tailored to the types of electronic components examined in CIRC-UITS. In response, the project will propose an enhancement to the repairability assessment aligned with EN 45554, shifting the focus toward the manufacturer's perspective, addressing critical factors like design for repair and optimized repair logistics, as the current standard is more end-user focused. The EN 45554 is a widely used norm to assess repairability potential of physical products. It is for instance the basis for the right to repair from the European Commission [12]. The norm was created from the consumer perspective, enabling consumers to make better decisions when buying products. Thus, pre-defined criteria cover aspects of the consumers concerns like the availability of repair manuals and the necessary working environments. However, the norm should also be able to be applied from the producer perspective. This can be challenging because some of the pre-defined criteria are not applicable for them. Hence, some of the criteria need to be redefined of new criteria need to be added. Also, when defining the fulfilment degrees for the criteria, standards of the industry need to be considered. This includes legally regulated spare parts stocking periods.

The norm promises to help companies understand how much better their new designs are in terms of repairability. This can be used internally to justify R&D expenditure and to show the progress of the design changes. Having a number instead of a rough description of the changes can help people without technical expertise to understand the effort. Furthermore, redefining new criteria also gives the possibility for companies to create scenarios that are linked to business cases. This way they can make a scenario analysis and build a path for improving their product for better repairability for the future.

The norm also has a disadvantage when producers are aiming to compare themselves with products that fulfil the same purpose. Since the norm needs to be defined for each product category, the typical parts of this product, their relative importance and assessment criteria need to be defined before the assessment itself can be carried out. This makes the norm inflexible for comparing e.g. filter coffee machines

Table 3 Pilot preferences on advisory services for assessments

PILOT	Predict Environmental Impact	Handle Uncertainties in Inventory Data	Find Best customization	Chatbot: a Textual Assistant for Result Interpretation	Enhancing Product Design and Comparison
Pilot 1	8	4	7	8	8
Pilot 2	5	3	5	6	6
Pilot 3	1	0	3	3	3
Pilot 4				2	
	14	7	15	19	17

and french presses. Both products fulfil the same need—making coffee—but with completely different parts. The norm would have to be defined for both products individually and thus, the results of the assessment would not be comparable anymore. Furthermore, rating the fulfilment of the criteria is highly subjective, making the results of different assessments (even for one product category) hard to compare. Having a third party carrying out the assessment would decrease this risk as i-fix-it does it for e.g. laptops and smartphones.[2]

In summary, the norm needs some adjustment for the companies to use them directly. It can be however very helpful in showcasing the own efforts in increasing repairability in comparison with old product versions or competitors with the same functionality. It is advantageous that the norm is already widely used, which increases the acceptance and the ease of interpreting the results. The norm has the limitations that applying the norm is subjective and no two different products can be compared directly.

4 AI-Based Advisory Solutions to Support Eco-Design, LCA and Recycling Assessment

For each pilot, specific advisory needs were identified; corresponding AI-based use cases were defined to address those needs. Table 3 reports the use cases listed in the first row and the pilots in the first column. Based on the ratings assigned to each need linked to a use case, the final score is presented. The focus will be placed on the highest-scoring use cases, briefly describing them in Table 3.

AI Techniques to Predict Environmental Impacts

Conducting a Life Cycle Assessment (LCA) study is time-consuming and requires detailed analysis. AI, particularly machine learning (ML), can be used to estimate the environmental impacts of products quickly and accurately. Various ML methods,

[2] https://www.ifixit.com.

including Artificial Neural Networks (ANN) and algorithms like multiple linear regression and decision tree regression, are employed to optimize LCA and predict outcomes [14]. A new method leverages Environmental Product Declarations (EPDs) to predict environmental impacts in four categories: global warming, fossil resource depletion, acidification, and ozone creation. EPD data is collected via automated web scraping, pre-processed with Natural Language Processing (NLP), and fed into tree-based algorithms, primarily Random Forest regression, to ensure high prediction accuracy. This approach optimizes model performance and avoids overfitting, with hyperparameters fine-tuned through grid search and cross-validation [13].

Best Design

In production processes, managing extensive Life Cycle Inventory (LCI) data is a constant challenge. When designing a product, numerous variables (design choices involving both product, manufacturing and supply chain features) can influence sustainability impacts, making it difficult to identify which variables are most significant. This challenge is compounded by various business and technical constraints related to production processes.

Having an advisory system to assist in finding the optimal configuration of variables, while balancing sustainability impacts and costs, would be highly valuable. The traditional approach to solving these problems involves linear programming and optimization of an objective function or brute force techniques, which can be resource intensive. In this context, AI can play a crucial role.

Chatbot

A chatbot is a software program designed to simulate conversation with human users, either via text or voice. Using artificial intelligence techniques such as natural language processing and machine learning, chatbots interpret and respond to user queries in a contextual and relevant manner. Some possible applications are: (i) Results Interpretation—If the LCA includes different scenarios or alternative products, the chatbot can help compare them, highlighting key differences and suggesting which option might be more sustainable; (ii) Assist the user—The chatbot can guide users through the different aspects of the LCA report explaining the significance of each section and how various data influence the outcome; FAQ Feature—by implementing an FAQ feature, the chatbot can provide quick responses to common questions about the LCA and its results (and/or on the S&C assessment and advisory tools) or guide the user during the utilisation of the LCA tool.

Enhancing Product Design and Comparison

A study developed a tool to support decision-making in the early stages of environmentally conscious product design. The tool employs an Artificial Neural Network (ANN) to analyse product lifecycle parameters (e.g., size, materials, processes, recyclability) and provide estimates of carbon footprint and lifecycle costs. The ANN is integrated into a Graphical User Interface (GUI) to enable real-time evaluation of design options. Effectiveness and efficiency were validated by comparing the tool's outputs with results from GaBi 9.2, a standard LCA software. Key steps include data collection, ANN training, GUI integration, and evaluation against traditional methods [14].

AI for Recycling

MARAS has a long history in developing and applying AI and neural net/fuzzy based approaches in process simulation and metallurgical plant optimisation [9, 10, 15, 16, 27].

Within the European Union 6th framework project SuperLightCar fuzzy recycling models to provide real-time design-related recycling calculations have been uniquely developed and applied, capturing the know-how from the complex simulation models [16]. Also, the processing, completion and structurization of design data being input to recycling and exergetic simulation assessment is extremely time consuming, as for this moment, this can only be done manually. Automation/digitalisation of input data is essential so that it can link the product design easily to a digital twin (simulation model) of a metallurgical and energy recovery processing infrastructure. Classification of input composition is part of this process in view of preparation of data sets from the detailed simulation models to create surrogate functions that twin the simulation model. The Recycling Simulation Models in HSC Sim are too complex to be integrated within the CIRC-UITS platform for advisory and assessment. Therefore, all input and output data from the simulation model, which calculates all flows, can estimate exergy dissipation but also environmental footprint information can be integrated into surrogate functions.

Based on the different designs, redesigns combined with the physics based versatile recycling flowsheet simulations in the recycling digital twin (Recycling Simulation Models), surrogate functions that twin the simulation model will be created. Neural net surrogate functions (AI) will be trained to translate the complex recycling simulation tools into easier to link and rapid calculating digital AI based tools as depicted in Fig. 4 these neural net—AI (Artificial Intelligence) based tools can integrated into design tools and the CIRC-UITs platform for use in for example design tools for rapid calculations.

5 Implementation of the Methodologies into Supporting Digital Tools

The LCS&CA customized framework based on pilot needs provide the methodological foundation for a set of digital tools developed within the project. Among them are GRETA, MARAS simulation, and a distributed Advisory service.

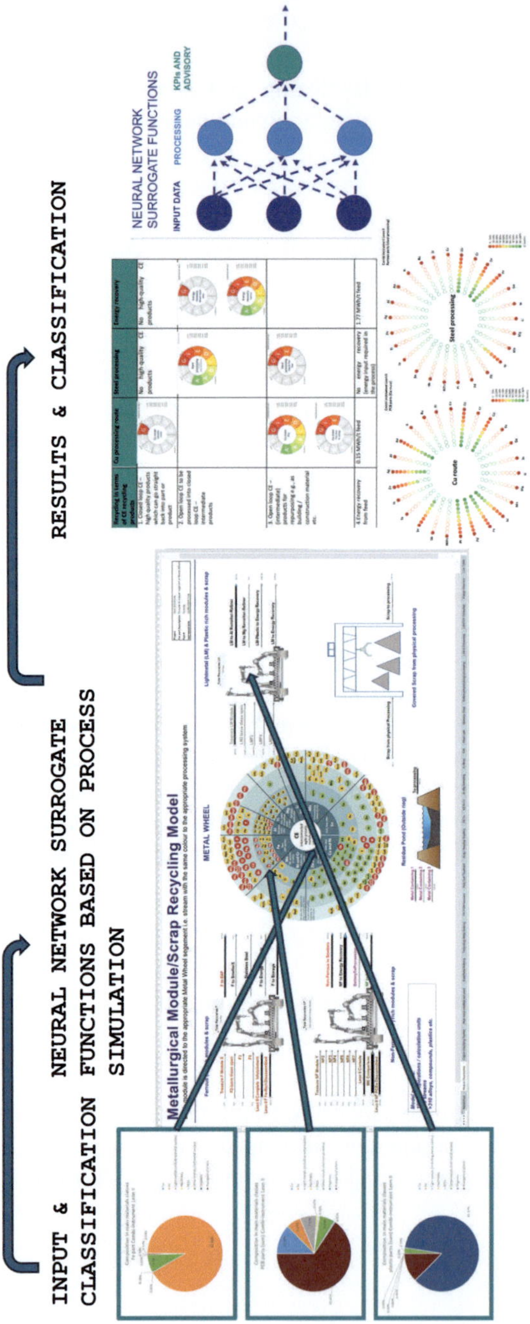

Fig. 4 Creation of neural net surrogate functions based on process simulations (MARAS)

5.1 GRETA

GRETA is a web-based application, developed with a microservices architecture, dedicated to assessing the sustainability and circularity of products and processes within the manufacturing sector. The tool is equipped with diagnostic and advisory functionalities, which empower users to enhance their production practices through data-driven insights. Created specifically with manufacturing companies in mind—especially those focused on sustainable, early-stage product design—GRETA allows users to simulate and compare different production scenarios even when limited data is available at the preliminary design stage. This simulation generates comprehensive sustainability profiles that balance environmental, economic, social, and circular factors, guiding users toward the most eco-friendly and cost-effective production options.

Within the CIRCUITS project, GRETA was expanded and adapted to address specific pilot needs and use cases. This included integrating CIRCUITS platforms with appropriate tools and technologies. GRETA's integration layer was designed to facilitate seamless interaction with a range of external components, functioning both as data sources and as recipients of GRETA's output. This integration spanned various types: A model editor for formalizing process and product models from a sustainability perspective, which could then be imported into GRETA; External data sources (such as dataspaces, blockchain solutions, databases, etc.), from which GRETA could gather inventory data or share sustainability indicators; Real-time sources (such as sensors, IoT devices, middleware, etc.) that provided dynamic data for specific parameters; External tools (such as ERP systems, CAD applications, etc.) that leveraged GRETA's assessment capabilities to enhance their own functions.

GRETA's sustainability assessment results, alongside additional valuable data such as scenario customizations, profile-specific parameters, and in-depth reports, were fully integrated into the Digital Twin (developed by OFFIS) and other external tools. This was achieved through the Data Manager sub-module within the Data Layer developed by TXT. By leveraging this layer, GRETA ensured that relevant data remained accessible and interoperable across multiple platforms, enabling real-time updates and synchronization. This connectivity supported the digital twin's accuracy and utility, allowing users to make informed, sustainable decisions with a holistic view of their manufacturing processes and their environmental impacts.

5.2 MARAS Recycling Process Simulation Models

Calculating true recycling rates in a product-centric manner is a cornerstone of the circular economy (CE).

The recycling process simulation models provide a rigorous and physics-based Digital Twin for the recycling processing system (see also [17, 18]). The systemic view and the detailed mass and energy balances these models provide, enable rigorous resource efficiency and sustainability evaluations for production and recycling processes and the systems they are a part of. The results obtained through simulation-based approach include environmental indicators, exergy, recycling and recovery rates, as well as the qualities and quantities of the recyclates, losses and emissions of materials during production recycling. The detail of the models provide the basis to define and improve environmental assessment, while at the same time providing product and recycling processing route specific EoL LCA input to LCA databases and to the LCA tools.

The digital simulation models can be linked to other tools for advisory, as it has to depth to distinguish differences in product designs with respect to recycling performance and environmental impact. It allows that all processing options in the recycling system, ranging from disassembly and sorting to metallurgical and other final treatment processes, are understood and optimally linked in fundamental detail and can be related to product design considerations. Advisory is defined on optimal flowsheet architectures, disassembly and modular design, repair balanced with recycling and Design for Recycling. Distributed Advisory Service.

5.3 Distributed Advisory Service

The distributed Advisory service operates across multiple standalone components within the project. Instead of a separate tool, advisory capabilities are embedded in components such as GRETA, MARAS Simulation, and the Digital Twin, collectively guiding decision-makers in the Eco-Design and End-of-Life stages. The following is a focus on these advisory functionalities, namely the Design Tool, the Decision Tool, the CRM Dashboard, the LCI Matrix Completion, the Repairability Assessment, The Recycling KPIs, the Metal Wheel, the Recycling Routes, the GRETA comparison:

The Design Tool is an interface of the Digital Twin through which the user, in this case the designer, specifies the product requirements. This feature, trained on previous assessment results, can provide instant feedback on certain KPIs without requiring a full assessment.

The Decision Tool is another feature of the Digital Twin. It supports the designer in making the final design choice by displaying a complete list of KPIs, including those related to repairability. In this case, the results are not provided in real-time, as all assessments must be completed first to obtain feedback.

The CRM Dashboard is an interactive decision-support tool that assists the designer in selecting raw materials by providing a criticality analysis of materials. Through visual indicators such as pie charts, maps, and radar charts, the dashboard displays detailed information on availability, import dependency, economic risks, and supply risks of raw materials. This data helps the designer make informed decisions by identifying the most critical raw materials and evaluating the impact of choices on project sustainability and supply chain resilience.

The LCI Matrix Completion, is an advanced feature developed to support LCA (Life Cycle Assessment) experts when managing an incomplete life cycle inventory

(LCI), where some essential product metrics are missing. Using artificial intelligence techniques, such as Monte Carlo simulation, the Matrix Completion analyses similar scenarios to generate plausible and consistent values based on the existing data, thereby filling in the gaps. This approach provides a more complete database, making the LCA assessment more accurate and reliable, even in the presence of missing information.

The Repairability Assessment Tool employs the methodology of the EN 45554 standard, which establishes criteria for assessing the ability to repair, reuse, and upgrade energy-related products. This tool allows the designer to calculate a product's repairability by inputting specific parameters, organized into categories, each with a weight. For each product component scores are assigned, which, combined with the weights, result in an overall repairability assessment.

The Recycling KPIs, advisory on optimal Recycling Routes and recommendations for Disassembly are all advisory functionalities relying on Recycling Simulation results and the evaluation thereof.

The Design for Recycling advisory based on the detailed process simulation models, which link in- and output flows of the recycling system, allow for pinpointing the hot-spots in Design with respect to recycling. The Metal Wheel provides a qualitative basis for this, while the quantified recycling KPIs and evaluation of recycling performance allows for unique, design specific DfR, by not only considering optimization of recycling rates, but also including achievement of high-quality recycling products (being indispensable for CE and reapplication of recycled materials in design) as well minimization of losses and emissions from recycling for both EoL products and components coming free from repair activities.

The Greta comparison is an advisory feature that allows designers to compare the sustainability impacts of alternative designs for the same product using radar charts. This graphical support relies on real-time LCA results calculated for each design. The system thus supports designers in making data-driven, informed decisions, guiding them toward the most sustainable design solutions.

Figure 5 illustrates the concept of AI-based distributed advisory services. Components marked as "new" indicate features developed within the project, while others are pre-existing and will only undergo testing. Subcomponents labelled with "AI" specifically denote those with AI functionalities. The connections represent the support provided to decision-makers, including eco-designers, end-of-life decision actors, and LCA experts.

5.4 AI-Based Distributed Advisory Services—Conceptual Framework

Figure 6 presents the high-level architecture of this system. Although distributed, the components must integrate and communicate via APIs. Additionally, the system will interface with external databases, such as Ecoinvent.

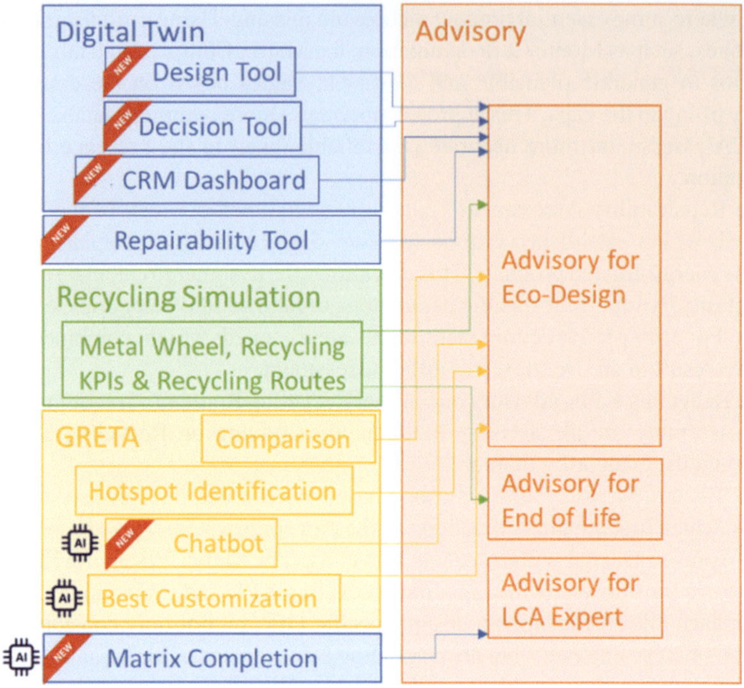

Fig. 5 AI-based distributed advisory services

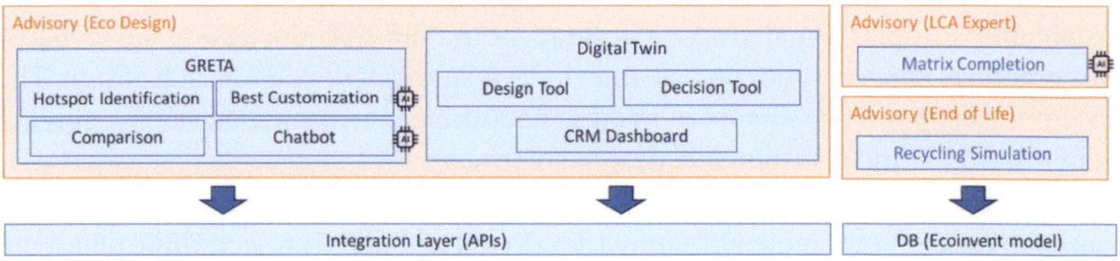

Fig. 6 AI-Based distributed advisory services—high level architecture

References

1. I. O. f. Standardization (2006) ISO 14044:2006 Environmental management—life cycle assessment—requirements and guidelines. ISO, Geneva
2. Damiani M, Ferrara N, Ardente F (2022) European Commission: Joint Research Centre. Understanding product environmental footprint and organisation environmental footprint methods. Publications Office of the European Union
3. I. O. f. Standardization (2017) ISO 15686-5:2017 Buildings and constructed assets—service life planning—Part 5: life cycle costing. ISO
4. SLA (2020) UNEP life cycle initiative. Guidelines for social life cycle assessment of products and organizations. United Nations Environment Programme (UNEP)
5. H2020. Grant agreement ID: 958448 (Online). https://circthread.com

6. H2020. Grant agreement ID: 722023 (Online). Available: https://www.treasureproject.eu
7. Maister K, Di Noi C, Ciroth A, Srocka M. PSILCA v3, GreenDelta
8. EM Foundation (2019) Circularity indicators: an approach to measuring circularity
9. Bracquené E, Dewulf W, Duflou JR (2020) Measuring the performance of more circular complex product supply chains. Resour Conserv Recycl
10. Wolf MA, Partl H, Wolf K (2020) The circular footprint formula (CFF) and its practical application
11. European Committee for Electrotechnical Standardization (CENELEC) (2021) EN 45554: General methods for the assessment of the ability to repair, reuse and upgrade energy-related products. CENELEC, Brussels, Belgium
12. E Parliament (2024) Directive (EU) 2024/1799 of the European Parliament and of the Council of 13 June 2024 on common rules promoting the repair of goods and amending Regulation (EU) 2017/2394 and Directives (EU) 2019/771 and (EU) 2020/1828
13. Koyamparambath A, Adibi N, Szablewski C, Adibi SA, Sonnemann G (2022) Implementing artificial intelligence techniques to predict environmental impacts: case of construction products,» Sustainability
14. Singh PK, Sarkar P (2022) An artificial neural network tool to support the decision making of designers for environmentally conscious product development
15. Bartie NJ, Cobos-Becerra YL, Fröhling M, Schlatmann R, Reuter MA (2021) The resources, exergetic and environmental footprint of the silicon photovoltaic circular economy: assessment and opportunities. Resour Conserv Recycl
16. Krinke S, van Schaik A, Reuter MA, Stichling J (2009) Recycling and DfR of multi-material vehicles (as part of 'Life cycle assessment and recycling of innovative multi-material applications' by). In: Proceedings of the international conference 'innovative developments for lightweight vehicle structures
17. Meskers C, Worrell E, Reuter MA (2023) Handbook of recycling: state-of-the-art for practitioners, analysts, and scientists, vol 2. Elsevier, The Netherlands
18. van Schaik A, Reuter MA (2024) Simulation-based design for recycling of car electronic modules as a function of disassembly strategies. Sustainability 16
19. van Schaik A, Reuter MA (2014) Material-Centric (Aluminium and Copper) and Product-Centric (Cars, WEEE, TV, Lamps, Batteries, Catalysts) Recycling and DfR Rules. In: Worrel E, Reuter MA (eds) Handbook of Recycling pp. 307–378 (updated in the 2nd Edition, 2023). https://doi.org/10.1016/B978-0-12-396459-5.00022-2
20. Abadías Llamas A, Valero Delgado A, Valero Capilla A, Torres Cuadra C, Hultgren M et al. (2018) Simulation-based exergy, thermo-economic and environmental footprint analysis of primary copper production. Miner Eng 131:51–65
21. Amini SH, Remmerswaal JAM, Castro MB, Reuter MA (2007) Quantifying the quality loss and resource efficiency of recycling by means of exergy analysis. J Clean Prod 15(10):907–913
22. Ignatenko O, Schaik A van, Reuter MA (2007) Exergy as a tool for evaluation of the resource efficiency of recycling systems. Miner Eng 20:862–874
23. HSC 10 (2024) Metso outotec research 1974–2022. www.mogroup.com
24. Reuter MA, van Schaik A, Gediga J (2015) Simulation-based design for resource efficiency of metal production and recycling systems. Cases: Copper production and recycling, eWaste (LED Lamps), Nickel pig iron. Int J Life Cycle Assess 20(5):671–693. https://doi.org/10.1007/s11367-015-0860-4
25. Handbook of Recycling, State-Of-the-art for Practitioners, Analysts, and Scientists, Second Edition (2023) Elsevier (Christina Meskers, Ernst Worrell and Markus A. Reuter) https://www.sciencedirect.com/book/9780323855143/handbook-of-recycling
26. Reck BK, Graedel TE (2012) Challenges in metal recycling. Science 337(6095):690–695
27. van Schaik A, Reuter MA (2007) The use of fuzzy rule models to link automotive design to recycling rate calculation. Miner Eng 20:875–890
28. Reuter MA, van Schaik A, Gutzmer J, Bartie N, Abadıas Llamas A (2019) Challenges of the circular economy: a material, metallurgical, and product design perspective. Annu Rev Mater Res 49(1):253–274. https://doi.org/10.1146/annurev-matsci-070218-010057

Ecodesign for ECUs

Karin Sämann and Achim Maat

Abstract Bosch Circular economy activities range from reusing products and their components to repairs and to remanufacturing, in each case with the objective of extending product and component life cycles. This chapter provides an overview of the Bosch Pilot (Pilot 1) and describe the product itself including lifetime and repair models and gives examples how eco-design can apply to the product. Starting with the outputs from the partner ideas and design are created to improve the environmental footprint of the product.

Keywords Electronic control unit · ECU eco-design

1 ECU

ECU means Electronic Control Unit. In this project the specific ECU is the Electronic Stability Program (ESP). The ESP makes a significant contribution to road safety by preventing vehicles from skidding, thus helping to prevent accidents and save lives. Vehicle manufacturers strive for personalization and differentiation, for example through driving dynamics and driving experience. This kind of product is mounted millionfold in cars all over the world. Remanufacturing and repair of used ESP devices for dedicated customers has already been established since some years at BOSCH in low volumes. BOSCH decided to focus on ECUs, to avoid cost increases due to low volume in the aftermarket or missing spare parts due to supply issues (e.g., chip crisis).

K. Sämann
Mobility Electronic, Robert Bosch GmbH, Reutlingen, Germany
e-mail: karin.saemann@de.bosch.com

A. Maat (✉)
Mobility Aftermarket, Robert Bosch GmbH, Hildesheim, Germany
e-mail: achim.maat@de.bosch.com

P. Rosa and S. Terzi (eds.), *Circular Approaches for the Ecodesign, Repair and Remanufacturing of Car and Mass Electronics*,
PoliMI SpringerBriefs, https://doi.org/10.1007/978-3-032-04301-6_5

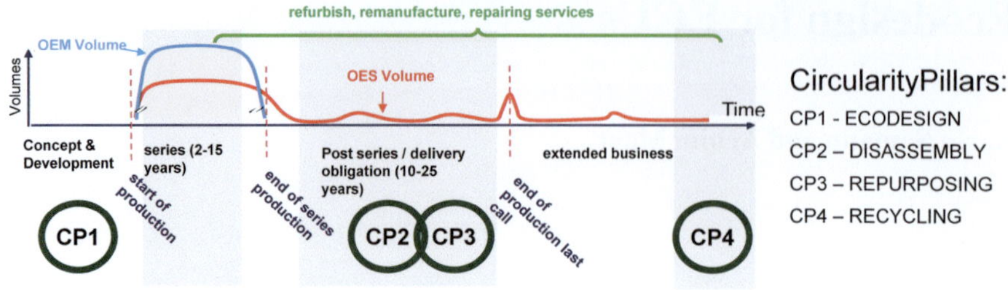

Fig. 1 Life cycle volumes of typica ECU

The product cycle of an automotive control system is about 10–25 years and includes a long-term supply guarantee also for control units and solutions for vintage cars (Fig. 1).

2 Eco-design

Eco-design is not only the availably of repair, it is a combination of activities to reach a competitive product with low CO_2 footprint, long live time and the best performance to go into a circular economy.

In the first step of Eco-design it is necessary to know about the "Status Quo" of the product. The "Status Quo" is the current design and depends on the materials and technologies used for the product. ECUs are in principle build up like the following graph. Main parts are a substrate with electronic components fixed on a base plate and packed in a housing (Fig. 2).

Within Circ-uits we shared the used technology and materials with specific partners to evaluate the Bosch ECU. The partners evaluated the ECU regarding hotspots,

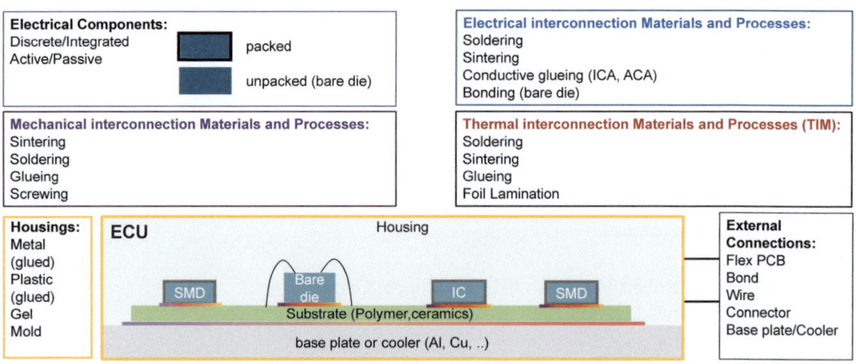

Fig. 2 Schematical build-up of an ECU including the different possibilities of interfaces

critical row materials (Offis), CO_2 footprint (Supsi), exergy (Maras) and recyclability (Polimi). You find the results in their chapters.

Each material and technology has specific properties and influence the reliability and usage time of the product.

Priority to reduce the CO_2 footprint of a product is to reach a longer lifetime. That means that a product performs as long as possible and can be repaired easily. For ECUs the product live target is about 10–25 years. Most of the ECUs are depending on the lifetime of the car, means the ECU is still working but the car is not. This long lifetime is due to very high-quality standards, robust design and smart development. Very often robust design stays in competition to repair or Eco-Design. Eco-design and quality must never contradict each other. This needs to be considered when we change the design or using recycled materials.

3 Implementation

3.1 Repairability

For those ECUs which are failing repair is an important option to extend lifetime.

In 1986, Bosch started to repair multimedia- and other electronics from cars and is extending the portfolio as needed since that time. To enhance this, the bridge to product design (eco-design) must also be built. The continuous improvement of the design with clear strategies for repair is important and will strengthen the position in the market and maintain or create jobs. Costs are and will remain an important factor, which is why it is important to set the course for effective repairs as early as possible in the development phase. The tool developed in the project is intended to support this. The experience that Bosch has built up over the last 20 years is anchored in the tool, concretized with KPIs and is then available to the developer. This enables us to ensure that knowledge is retained and passed on.

For the ESP different trails for opening and repair were performed. The design elements have a big impact on the repairability of the product (Table 1).

Eco-design plays a big role and need to be considered in the very early state of a product design. With an Eco-design for repair the possibility of repairing is given as well as the possibility of reusing of the components and parts of the ECU.

3.2 Reusability of Surface Mounted Devices (Automatization of Dismantle)

To ensure competitiveness, the repair needs to be as efficient as possible. In cooperation with Politecnico di Milano the automatization of component resoldering from the PCB is developed.

Table 1 Dismantling steps of EXP9

Opening Laser-welded plastic The Cover is welded to the housing by laser transmission. Removing the cover is a destructive process. A further use of the cover is not possible. A reclosing is then challenging when no original cover is available. To meet the technical requirements with any kind of cover needs to be specified and tested	
Releasing Pressfit technology Press-fit technology connects the contact terminal and external solenoids with the main printed circuit board (PCB). The press fit pins have an elastic press-in zone with multisprings. The PCB is designed with plated through holes. (PTH) When PCB is pressed into the press-fit pin, each pin is getting deformed by insertion. A permanent contact normal force is achieved. The tin plating in the press-fit-pin-system supports additionally a cold-welded interconnection Both mechanisms (springforce and cold-welded interconnection) make it difficult to release the PCB from the press-fit connection. There is a high risk of microcracks in soldered components or in the PCB itself when you release the PCB from this strong connection There is no experience in reuse the pressfit connection a second time with the same pins and the same PCB. To meet the technical requirements needs to be specified and tested	 

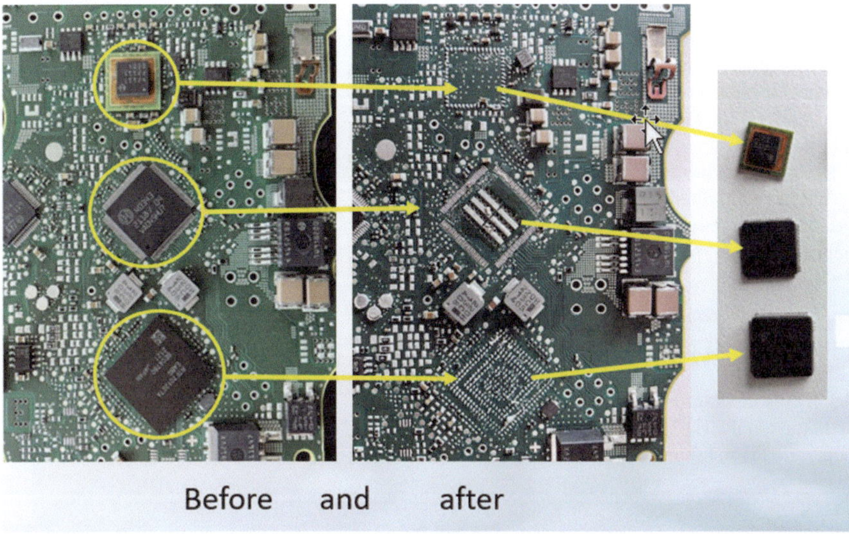

Fig. 3 Dismantling of the three most valuable components

Therefore, we send PCBs to our partner to evaluate the automatic dismantling. After dismantling we will test the chips again (see results of Politecnico di Milano) (Fig. 3).

3.3 Mechanical Recycling (Shredding, Sorting)

We have synergies with repair in recycling, as efficient and good quality recycling works best when the materials can be separated from each other as individually as possible. It must be ensured that disassembly is simple, cost-effective and, if possible, sorted by type.

At the end Eco-design also must consider the possibility of a high recycling rate. Because every product will have finally an end. It must be ensured that at the end the materials are not "lost". Here two aspects due to the results of Erion are important.

1. Avoiding material combination that makes recycling impossible: for example, strong connections between plastic and silicon parts results in poor recycling quality of the plastic.
2. Design the product in a way the separation of pure material is as easy as possible.

For the ESP different paths for mechanical recycling were performed. The design elements had a big impact on the mechanical recycling of the product (Table 2).

Table 2 Eco design option for plastic recycling

Silicone filler for connector pins Connector pins are assembled to the housing by injection-molded thermoplastics. Pins and connector are filled by silicon rubber to ensure technical requirements. Silicon and plastic relate to high adhesion. Separation is not possible by shredding	
CIPG Sealing (Cured In Place Gasket) A CIPG is the interface sealing between actuator and the ESP Separation is not possible by shredding	
Silicon balls (Pumpers) Clipped silicon pumpers are used for mechanical compensation in the product. These balls can be released easily. Separation during shredding is possible. The red color enables a good visual detection and therefor sorting of these balls	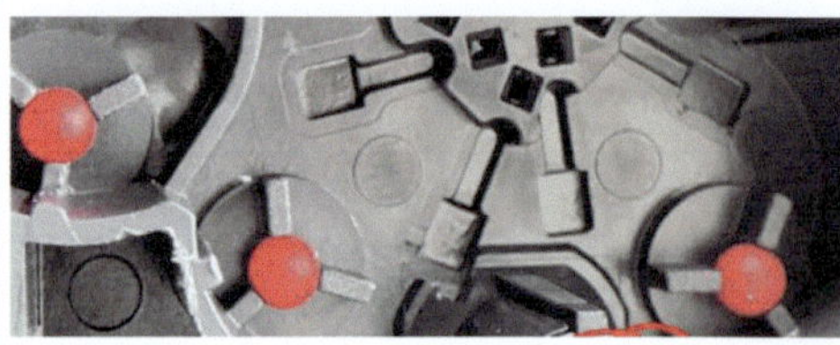

3.4 *Chemical Recyclability (Smelting)*

Most of broken parts will end in any kind of smelting process. They will neither be separated from vehicles nor dismantled or sorted. Melting means to recover the product exergy in form of alloy, slag and heating.

MARAS analyzed the recyclability of our product in total and of all components that are receive after one or more dismantle steps. This is done based on models that are linked to real melting processes in the recycling industry (Refer to WP3).

The received recyclability rate in form of an efficiency value of our ECU is low. Only 8.8% of the exergy in the product can be recovered in usable alloys. Dismantling the solenoids and melting them will improve the recovered exergy to 22.9%.

The body is more or less "fuel that is burned in the smelter". The efficiency for recover alloys is only 3.6%, due to its high quantity of inorganics. On the other hand, the body and the inserted PCB is responsible for > 90% of the total CO_2 footprint of the product. From sustainable point of view a further or extended use of the PCB or components of the PCB makes therefore more sense (Fig. 4).

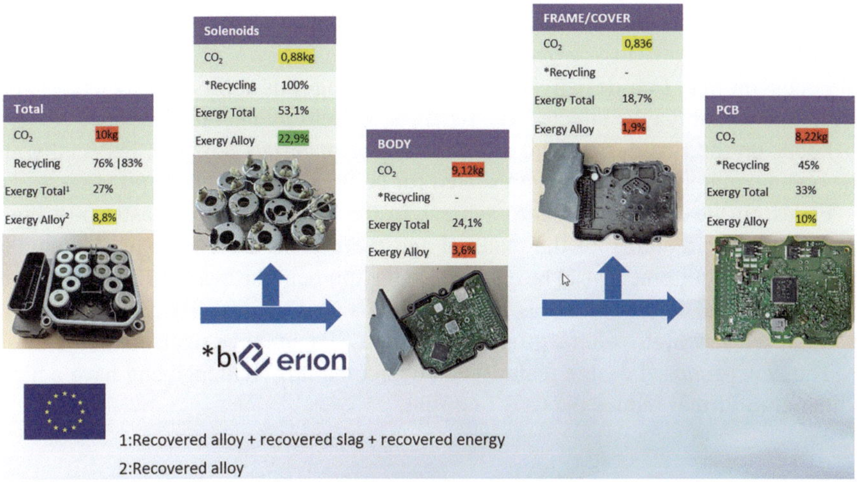

Fig. 4 Evaluation results of ESP9 regarding CO_2, exergy, and recycling

3.5 Materials

New materials provided by the project partners Alpha enable more energy-efficient production of control units. This means that energy, costs, and the associated CO_2 can be saved. The energy costs for solder pastes can be reduced by 5%, both in terms of production and the reverse process. Alpha tested already the solder paste and the performance (see Alpha results). Unfortunately, the performance is not fitting to automotive standards.

Recycled plastic material for technical applications are already available on the marked, they are mostly not 100% recycled but up to 80% is available. The rest is virgin material to ensure that the mechanical properties are not chancing too much. For recycled plastic instead of virgin material you can save half of the CO_2 footprint per Kg. The calculation with recycled materials you can see in the result of Supsi.

4 Result

4.1 Eco-design Features

In cooperating with the partners in the CIRC-UITS project we defined the following design features for a new kind of ECU (Eco-design). It supports repairability, recyclability and reusability of valuable components. The product (technology and materials) itself remains nearly untouched to keep all the benefits of an already CO_2

optimized product. Each feature can have an impact to more than one sustainable pillar (Table 3).

Focus on the Eco-design is the disassembling possibility for reaching the components that maybe effected in a repair or are reusable in other products.

Focus on the Eco-design is to make sorting of materials possible. Especially all the organic materials should be prevented for being lost in a melting process.

Focus on the Eco-Design is to achieve in the end a recycled plastic of a high quality. Then circularity of the material and the use in new Bosch products with high quality requirements can be ensured. In the moment the availability of recycled PBT30 material in the market is low in quantity and quality. Therefore, recycled material for new high-volume products like the ESP cannot be realized today.

The new proposed design features can overcome this challenge and have a high potential to further reduce the CO_2 footprint.

Table 3 Eco design and the effect on the 3 Rs (repair, reuse, recycle)

Change	Result	Repair	Recycle	Reuse
Dual use laser welded cover. Cover can be used for a second time	Achieves a second area, where cover and housing can be assembled again after a destructive opening procedure. Glueing or welding is possible in a repair process	x		
Additional housing element for pressfit release process	A housing element stabilize the PCB when lifting from pressfit pins. It spreads the high force to allowed areas. Pins and PCB are protected against mechanical damage during releasing process. An insertion of a repaired PCB after is possible. An additional fixation by soldering is then still necessary	x		x
Red colored solid gasket sealing	PBT30 GF can be separate from the sealing. Sealing can be identified by red color. For repair it is easy to exchange or even be reused for other ECUs	x	x	x
Remove Silicon fillers at connector	Can be removed when it is ensured that the vehicle plug works as specified. Silicon filler is just a backup. PBT30 GF can be separate from the sealing. (For our product the environment specification must be changed)	x	x	x

Eco-design for Tire Pressure Monitoring Systems

Philippe Lopez, Stéphane Couarraze, and Claudio Travi

Abstract To test the toolbox link to CIRC-UITS project, a next generation of sustainable automotive product has been set up with the TPMS (Tire Pressure Monitoring Sensor) including challenging technologies leading to repairable, recyclable and reuse.

Keywords TPMS · LCA · Circularity · Eco-conception · Recyclability · Recyclable · Reuse

1 Introduction

Sustainability is a key factor leading to target achievement from the EU Green deal by 2040. Within the Automotive industries the use of resources and especially the use circular economy is a very challenging topics, starting with the eco conception of the product, then with the new business model obliging also a new way of thinking for manufacturing. Within the development of TPMS (Tire Pressure Monitoring Sensor) next generation of product we will see how to save not only by using sustainable material but also thinking forward, by using new concept helping the full recovery of the different parts, including electronics, soldering process, battery and PCB components. These eco-design concepts will also allow repeatability, extending the life cycle ready to be used on new circular business models and manufacturing.

The application of the new design will lead to new ecological performances within the LCA (Life Cycle assessment).

P. Lopez (✉) · S. Couarraze
AUMOVIO France SAS—Tire Information System Unit, Toulouse, France
e-mail: Philippe.lopez@aumovio.com

S. Couarraze
e-mail: stephane.couarraze@aumovio.com

C. Travi
MacDermid Alpha, Langenfeld, Germany
e-mail: Claudio.Travi@macdermidalpha.com

P. Rosa and S. Terzi (eds.), *Circular Approaches for the Ecodesign, Repair and Remanufacturing of Car and Mass Electronics*,
PoliMI SpringerBriefs, https://doi.org/10.1007/978-3-032-04301-6_6

The tools box developed within CIRCUIT are helping the different design choices according to the different sustainable performances expected.

Within this project we've developed several TPMS concepts based on sustainability targets, all based on completely new technologies (Out of the state of the arts). We've evaluated them within the CIRC-UITS tools box, checked their performances, made choices, accordingly, set up prototypes and tested them within an automotive environment. (From Linear development taking into consideration quality, price and time to Circular life cycle, taking in additional sustainability in terms of business, product strategy and design.)

2 Pilot Plant Design and Description by Eco-design Evaluation

The overall pilot project consists of developing several technological solutions, using the CIRC-UIT toolbox, and then evaluating the best technological solutions. Once is done, set up prototypes and test them within automotive environmental conditions (Fig. 1).

The target expected was, starting for the high runner TPMS currently in production, what are the new performances expected:

- Reduction of Co_2 eq. compared to the current high runner in production, using sustainable materials (Target—20%).
- Product fully disassembled and repairable.
- Using sustainable materials and able to be fully recycled.

To develop new technologies and evaluate the performance we've set up a process flow to achieve the target expected (Fig. 2).

Based on this process flow of development, we set up a planning (Fig. 3).

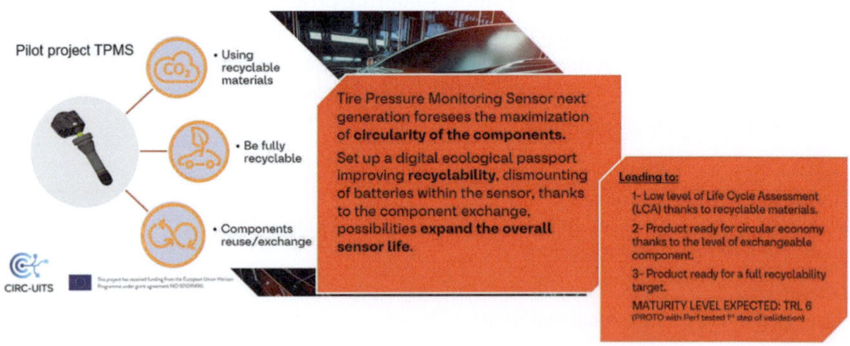

Fig. 1 TPMS product targets

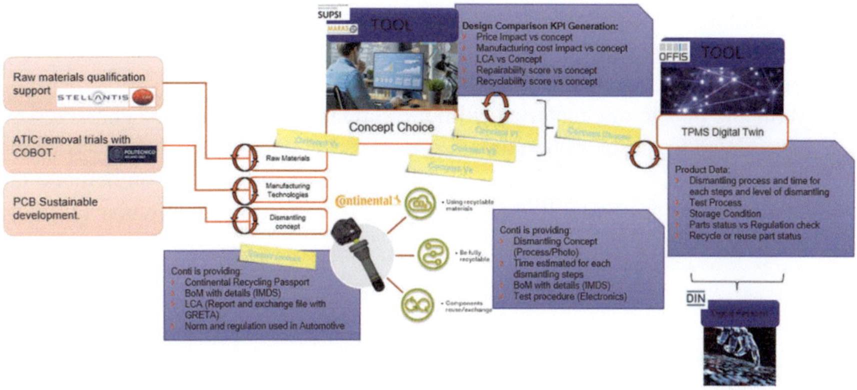

Fig. 2 Process flow and interaction mapping within the CIRC-UITS eco system

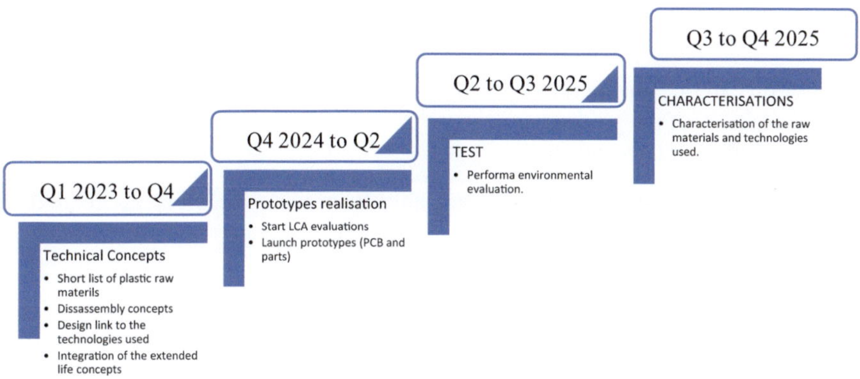

Fig. 3 Development process flow and planning

3 Performances Evaluations

3.1 Technologies Developed

Within the pilot, we've developed 6 technologies and validated 2 technologies of PCB sustainable within 2 prototypes.

6 Technologies out of the state of the art:

HW component recovery concept, thanks to specific technologies allowed with the support of MacDermid Alpha supports us on low temperature soldering past and soldering processes.

Battery exchangeable, the battery even with high performances of the connections is now exchangeable, increasing the life cycle of the product, and making the battery recyclable.

2 Technologies of PCB sustainable are tested within the product specification, making the PCB recyclable.

Rubber valves is sustainable.

Sustainable plastic raw materials with the support of Stellantis and characterization tests related to plastic materials.

Development of concepts of ECO-DESIGN able to disassembly on demand the product allowing repair and full recycling of the product. Allowing circular economy business model.

3.2 Sustainability

Once these technologies bricks are developed with 2 concepts (Cage and badge) the sustainability evaluation, by choosing the best human/machine interface to be used within CIRC-UITS toolbox and checking the different performances of the concepts like:

- LCA on carbon equivalent.
- Recyclability of the product
- Repairability of the product.

Set up a digital twin of the product able to provide the elements related to disassembly, repair, and test the product.

4 Conclusion

CIRC-UITS project provides us a huge opportunity to develop and diffuse within Continental the advantage of the ECO CONCEPTION and the TOOLS related to evaluate the performance of the product during the development. The technologies developed for the prototypes and the process set up for the toolbox are not only for TPMS applications, but for any products within Continental. Allowing us perspectives for the next step: Circular economy business model with product ready using eco-design.

Eco-Design for In-Mold Electronics

Stephan Harkema, Maarten Bakker, and Corné Rentrop

Abstract In-Mold Electronics (IME) is an attractive technology platform for smart surfaces based on printed electronics. Electronic devices are manufactured on and fully embedded within thermoplastics to protect the electronic functionalities from external influences. IME devices for automotive and household applications are on the verge of mass production. Any early-stage considerations regarding the environmental impact of IME, potential improvements to its sustainability and possible circular strategies that may technologically be feasible may enable designers and producers to incorporate suitable eco-design measures that will save costs and reduce the overall environmental impact. Moreover, they may enable complying with existing and/or upcoming EU legislation targeting consumer rights and supply chain independence through efficient recycling and effective repairing. In this Chapter, we will discuss the IME technology and recent scientific results obtained in the EU CIRC-uits project on lifetime extension of IME. The introduction to IME will continue with a lifecycle assessment to pinpoint the major contributors to the environmental impact. In the framework of EU CIRC-uits, we extended the design-for-recycling principles explored in previous and existing EU projects Treasure and Unicorn to achieve repairability of IME. In addition to the technical feasibility of repairing, we determined that repairing is environmentally less impactful than replacing a defective device, even when using incineration. Recycling of plastics and metals would, however, greatly contribute to a further reduction of the environmental impact.

Keywords Eco-design · In-mold electronics · Circular economy · Circular strategies

S. Harkema (✉) · M. Bakker
TNO at Holst Centre, Eindhoven, The Netherlands
e-mail: stephan.harkema@tno.nl

M. Bakker
e-mail: m.bakker@tno.nl

C. Rentrop
TracXon, Eindhoven, The Netherlands
e-mail: corne.rentrop@tracxon.tech

P. Rosa and S. Terzi (eds.), *Circular Approaches for the Ecodesign, Repair and Remanufacturing of Car and Mass Electronics*,
PoliMI SpringerBriefs, https://doi.org/10.1007/978-3-032-04301-6_7

1 Introduction

Scientific evidence that printed electronics is a suitable green alternative to printed circuitry boards is increasing [7, 12, 14, 27–30, 33, 37, 43, 41, 46]. In some versions, the likeness of printed electronics applications to printed circuitry boards is quite striking, as multilayer metal circuitry is combined with discrete semiconductor components, such as chips, resistors, light-emitting diodes, capacitors, and so on. There are significant differences, however. The printed circuitry boards, or PCBs, have pure metal circuitries realized by plating or lamination of copper films and are patterned to form the circuitry elements by wet chemical etching. The substrate is typically based on epoxy with glass-fibers, also known as FR4 [15, 33, 35], or flexible polyimide in what are called the flexPCBs. In printed electronics, PET, PU and PC are most typically used instead. These materials are not thermosets, but thermoplastics, meaning that these materials become soft again at elevated temperatures. This is beneficial for processibility, but also for recyclability. Epoxies crosslink under i.e. chemical reaction or UV treatment and remain stiff when heated. This is beneficial for certain applications that need to tolerate higher temperatures, but not ideal from a recycling point of view. Moreover, the various thermoplastic substrates used in printed electronics are thin and often transparent, more flexible, formable and comfortable on the body, enabling a plethora of new applications.

In-Mold Electronics, or IME, is an attractive alternative for conventional electronics based on printed circuitry boards (PCBs) for e.g. domestic appliances and automotive due to its form-factor, light weight, seamless design, diversity in functionalities, and high level of integration. IME is a version of printed electronics in which a formable thermoplastic, often polycarbonate, is chosen as a substrate onto which the circuitry and semiconductor components are applied. High pressure forming of this functional substrate creates a custom shape, potentially unique for each application. On one or both sides of the substrate, additional layers of one or more types of plastic are applied, e.g. PC, PC/ABS, PUR, PMMA, by means of injection molding, among others. This fully embeds and protects the circuitry and components. In-mold electronics may also be realized by two separate films, one for decorative and the other for functional purposes. Injection molding of the encapsulating resin or resins is done in between the two films or foils, thereby bonding these together.

In this chapter, we will describe the application of In-Mold Electronics in an automotive mid-console unit as an example of its possibilities. Subsequently, the IME developments and goals are described in the framework of EU project CIRCuits. Following this technological introduction in Sect. 2, the so-called hotspots from a recent life cycle assessment are described in Sect. 3. Section 4 focuses on the potential circular pathways towards reducing the environmental impact of IME from a material and production point of view, including material circularity and repairing of IME. Section 5 offers concluding remarks.

2 IME Devices: Composition and Manufacturing

Like many other electronics and electronic devices, IME combines plastics with metals, semiconductor technology and coatings that improve aesthetics or provide a function within the layer stacking for protection, adhesion, sensing or alike. The coatings and plastics are chosen to provide a highly reliable part that can last for years. Metals and components are largely embedded within these plastics. While this is highly favorable for protecting the electronic functionalities, this is less than favorable for recycling at end-of-life.

The CEN workshop document CWA 18119:2024 by the EU Treasure project [6] provided proposals for effective end-of-life recycling of automotive parts, including IME. It also provided a general composition of IME. IME largely comprises thermoplastics, such as polycarbonate (PC) and Acrylonitrile Butadiene Styrene (ABS). Silver is printed as a metal ink to provide functionalities as circuitry and sensors, and as conductive adhesive for semiconductor components. Silver is typically present in a low concentration, often in the range of 0–1 w-%. Other strategic and/or critical metals may be present in lower amounts in surface-mounted devices. Polycarbonate, ABS and possibly polyurethane make up a total amount of 95–98%. The rest of the polymers are used to create binders, adhesives, graphic layers and/or thin anti-scratch layers.

This chapter is dedicated to one example of IME in an application, in this case a prototype automotive mid-console, made by Holst Centre, as shown in Fig. 1. This device was manufactured using a single substrate, which is one of the possible approaches for IME. IME devices may be realized with one or two substrates and with one or more consecutive injection-molding steps, see for instance a recent review on In-Mold Electronics by Beltrão et al. from 2022 [4]. In the case of the two-substrate approach, the bulk plastics are injection molded in between the two substrates where one is functional and the other serves as decorative layer. In a one-foil approach, the decorative exterior may also be realized by an injection-molding step of e.g. polyurethane (PUR). The encapsulation of the electronics is injection molded onto the electronics and may also incorporate features or segments to improve the outcoupling of light created within the device using light emitting diodes. One example approach is provided in Fig. 2 from C. Goument et al. [17]. The IME device Fig. 1 was made in similar fashion.

The IME mid-console panel in Fig. 1, that serves as the example in this chapter, was manufactured using sheet-to-sheet processing onto a 500-micron-thick polycarbonate substrate of 390 $\times$ 260 mm^2 (Makrofol DE 1-1, Covestro), onto which the following layers were screen printed: (i) Ag fiducials for subsequent aligned printing steps, (ii) 3 layers of black, (iii) 3 layers of white graphic inks (Noriphan N2K 945 and 954, Proell), (iv) a first layer of Ag (DuPont ME603), (v) three layers of dielectric (DuPont ME779) and (vi) a second layer of Ag. All layers were cured using a convection oven set at 80 °C for the graphic layers (10 min), 120 °C for the Ag layers (10 min) and dielectric (20 min). Post-curing of the graphic stack, before application of the Ag circuitry, occurred for 2 h at the same temperature of 80 °C. Components

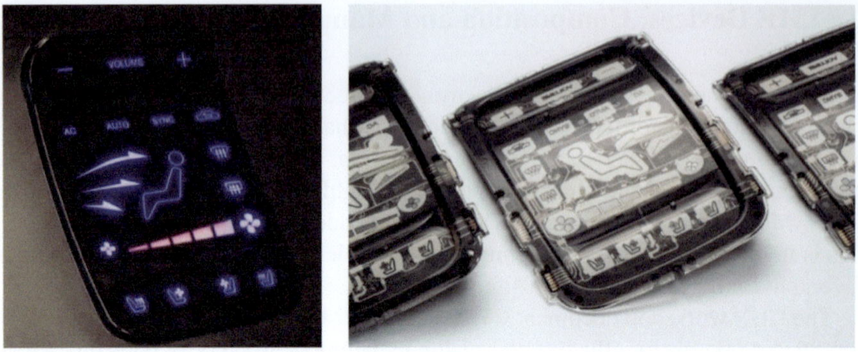

a) b)

Fig. 1 IME mid-console touchscreen, as shown in our LCA study from the front in its on-state (**a**) and back (**b**)

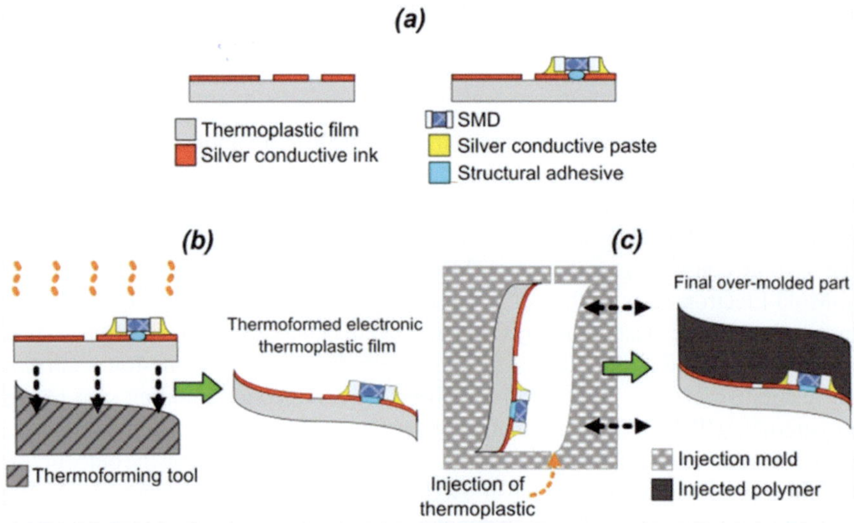

Fig. 2 Example IME production process (from: [17])

were applied onto the graphic layer with a conductive adhesive using a Mycronic My200DX-14 Pick and Place machine. A total of 134 components were applied: (i) 2 Atmell AT42QT2120 Q-touch chips, (ii) 26 RHPE 0402-IMP10k/47 k resistors, (iii) 54 CHPE 0402-IMP capacitors, (iv) 52 RGB smart side LEDs SK6812side. The conductive adhesive was cured at 120 °C for 10 min. The structural adhesive (underfill) was allowed to creep underneath the components to provide stronger adhesion to the substrate and circuitry. The flat PC substrates with circuitry and components

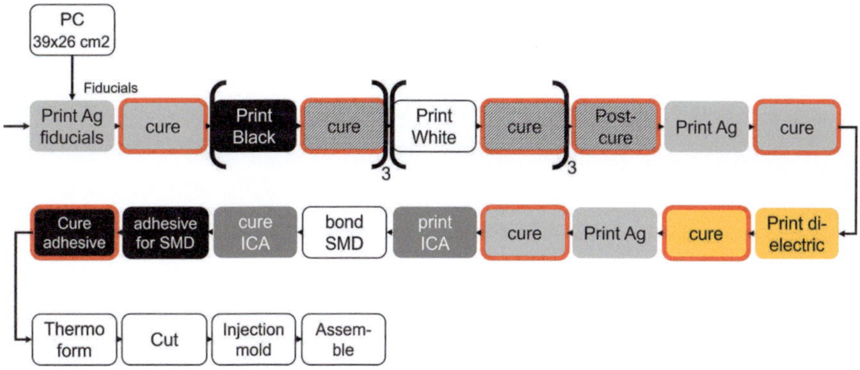

Fig. 3 Production steps for the IME mid-console prototype from Fig. 1

were thermoformed using a dedicated mold in a Niebling high-pressure thermoforming tool. After laser cutting and subsequent milling of the edges, the prototype was finalized by injection molding an estimated 200 g of transparent polycarbonate resin for samples with a transparent backside finish. Prototypes with advanced light management solutions were developed separately. Light guides were applied onto the substrate and LEDs before injection molding to provide confinement of emitted light and improve light outcoupling efficiency. For the LCA calculations of this device, described later in this chapter, the light guides are not treated separately and are considered indiscriminate from the rest of the injected resin. The key steps to produce the IME device are shown in Fig. 3.

Within the framework of EU CIRC-uits, IME is developed towards improved sustainability by (1) embedding all of the functionalities of the external printed circuitry board (PCB) and (2) repairability. PCBs are known to cause considerable environmental impact [15, 35, 41] and reducing the size or omitting the PCB as a whole is a strategy towards reducing the overall environmental impact of IME [25, 26]. Figure 4 shows the IME demonstrator that was developed by TNO at Holst Centre and TracXon. The external PCB was made redundant by embedding all functionalities within the device (touch sensing, light, gesture sensing, OLED (organic light-emitting diode) display and all driving electronics). At the time of writing, an LCA is in preparation by project partner SUPSI. The second point of repairability is addressed in this Chapter and focuses on demonstrating the technical feasibility of repairing IME, including an impression of the impact on the environment and costs.

3 Environmental Impact of In-Mold Electronics

For printed electronics in general, considerable scientific support the claim that printed electronics are more sustainable than their conventional PCB counterpart, however, IME parts contain a significantly larger amount of plastic. Various example

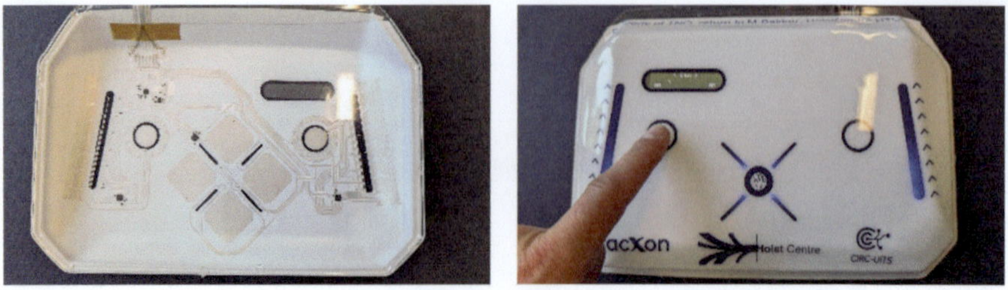

a) b)

Fig. 4 **a** backside of the EU CIRC-uits IME pilot demonstrator before laminating the OLED and further encapsulation, **b** frontside view showing light emission from LEDs as well as the OLED display and operation of one of the capacitive touch buttons

applications are e.g. 2–3 mm thick, instead of 100–250 microns, which are typical substrate thicknesses. The additional plastics add considerably to the environmental impact. In literature, few assessments are available that quantify the environmental impact for IME. In the next section, the life cycle assessment (LCA) of the IME automotive mid-console from Fig. 1 is elaborated on.

3.1 Life Cycle Assessment of an Automotive Mid-Console IME Panel

Godoi Bizarro et al. studied the environmental impact for the mid-console panel that combines an IME touchscreen and a PCB-based driving unit [16]. The aim of that study was to obtain a complete overview of the environmental impact of IME devices and their PCB counterparts from cradle-to-grave. Here, we only refer to the parts of that life cycle assessment that address the manufacturing and end-of-life to describe relevant environmental hotspots and to address eco-designing of IME in subsequent sections of this chapter.

Adapted results of the LCA, shown in Fig. 5, illustrate the contributions of raw materials and manufacturing of the IME part to the overall environmental impact. Quite clearly, polycarbonate provides a considerable contribution to several midpoint categories as the primary plastic in this plastic-rich part. Aside from the contribution of power consumption during production, Ag provides a significant contribution to multiple categories, including marine and freshwater ecotoxicity and mineral resource scarcity, but not so much to the global warming potential (GWP). A focus on eco-designing IME based on the GWP would thus underestimate the overall impact of Ag on the total environmental impact of this metal despite its minor weight contribution to the part. Major improvements may be expected when addressing the primary plastic, Ag and power consumption during production.

Regarding end-of-life, it was assumed by the authors that the IME and PCB would be separated at EoL and are disposed of in two different waste streams. The

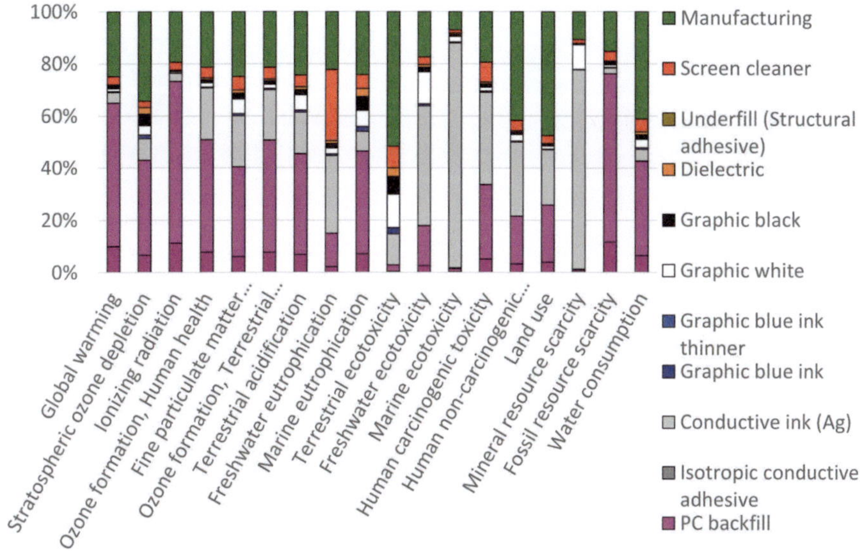

Fig. 5 Environmental impact of the IME mid-console in Fig. 1 consisting of contributions of raw materials and production

IME is expected to be incinerated for energy recovery while the PCB is disposed of as electronic waste. The metals from the PCB may be efficiently metallurgically recovered. Figure 6 provides an overview of the burdens and benefits of incinerating the IME part for the recovery of heat and electricity. The minor amount of Ag was not considered in this EoL calculation due to its small weight contribution to the part. Metallurgic recovery of the Ag from IME is environmentally desirable and possibly cost-effective due to the high pricing of Ag at this moment, however, is a challenge for IME devices due to the full encasing of all semiconductor components and printed Ag in the plastics.

The next section of this chapter continues with Eco-designing of IME for improved end-of-life material recovery and lifetime extension.

4 Eco-designing In-Mold Electronics for EoL Treatment and Extension

4.1 Improving Material Circularity

In the waste hierarchy for a circular economy [11, 13, 27], landfill and disposal without recovery of materials are least preferred, followed by incineration that at least provides benefits in the form of recovery of heat and electricity. Rethinking and

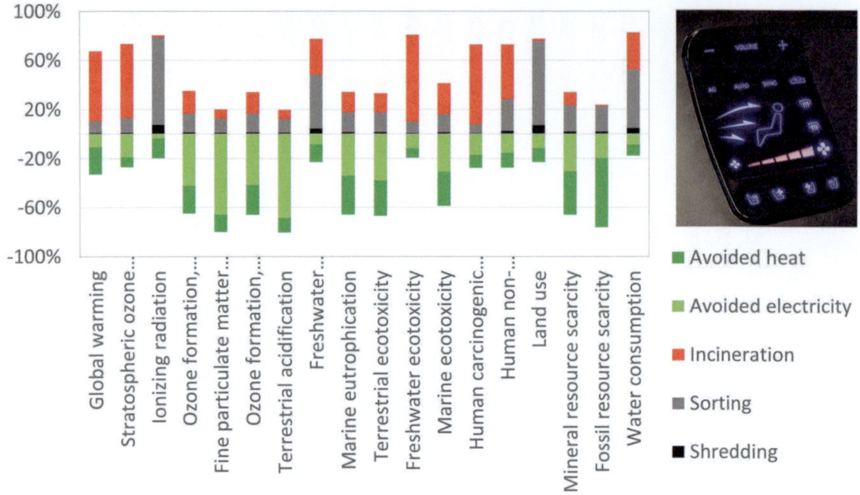

Fig. 6 Environmental impact of the incineration of the IME mid-console for the recovery of heat and electricity. Values > 0% are additional burdens/emissions, while values < 0% are benefits/ reduced emissions

refusing hazardous materials, impactful disposables, excessive packaging and low-lifetime products are at the top of the hierarchy [10, 13, 27]. In between, a circular economy is benefited by material circularity: recovery of materials at end-of-life, minimizing production scrap and alternative services and business cases, such as components as a service [21]. Preferred manufacturing techniques for IME in EU CIRC-uits and EU Unicorn involve digitizable additive manufacturing techniques by applying conductive layers only where needed, thereby avoiding the necessity to remove excess metals and the subsequent recycling thereof. Material circularity achieved by recovery and reuse of materials from H&PE devices, however, is not straightforward. Recycling of H&PE in existing WEEE (Waste from Electrical and Electronic Equipment) recycling plants may be quite challenging, surprisingly due to their unique selling points: their flexibility, light weight, low metal content and protection from environmental influences such as moisture. In traditional recycling plants, ferrous and high-metallic fractions are separable from other waste by magnetic and eddy current methods. PCBs are magnetizable and can be ejected by eddy current. H&PE remains part of the plastic waste stream due to the lack of detectable and magnetizable metal content. Within the plastic waste stream, H&PE forms a polluted fraction made of either PET, PC, TPU, ABS, PUR or even a combination of some of these, depending on the application. As a result, H&PE in the plastic waste stream will likely face being incinerated with recovery of heat and electricity instead of being recycled. For that, H&PE must first be recognized as metallized plastics, potentially also containing semiconductor components.

Like WEEE based on H&PE, recycling of automotive electronics based on In-Mold Electronics faces the same challenges. In CEN Workshop Agreement CWA

18119 [6], a similar end-of-life outcome was believed to be likely. Recent studies on the inclusion of design-for-recycling principles in IME [18, 5] offer a different outcome as these enable dismantling at end-of-life with the potential to improve the recycling rates for EoL of IME.

At present, if and when offered for recycling, IME is most likely incinerated at end-of-life for the recovery of heat and electricity. Closed-loop thermal mechanical recycling is not suitable for IME, as IME devices contain a significant number of pollutants, including precious metals (Ag), various graphic inks, possibly other plastics (ABS, PUR and/or PMMA), as well as semiconductor components. Other recycling methods may prove more useful, such as (smart) pyrolysis, and focus then on the liberation of metals rather than the recuperation of plastics [1]. Alternatively, plastics may be recovered by physical recycling via dissolution instead [3, 9, 32, 36, 44, 47]. How and how efficiently metals will be recovered from the IME device is a topic for further research.

The application of design-for-recycling principles has allowed project partners in EU project Treasure to target the Ag for recovery. Following necessary device separation to allow dissolution of the silver, the IME functional substrates were provided by TNO at Holst Centre to University of L'Aquila for hydrometallurgy [23, 40]. Public reports from the EU Treasure project provide further information on the recovery of silver [42]. A two-stage leaching process at laboratory scale yielded 85% dissolved Ag. Electrowinning of Ag from the solution was achieved at a yield of 87.5%, however, it was noted that at industrial scale the yield will be 95% or more. On the pilot scale, a yield of 81.2% was obtained for a two-stage leaching process. By optimizing the electrodeposition stage, and conducting pilot testing for each cycle's solution, an overall silver recovery of 97.5% was obtained for electrowinning. This efficiency was achieved with an energy consumption of 8.5 kWh/kg of recovered silver [24]. MARAS described in the same report the yields of metal recovery from IME parts using metallurgy at economy-of-scale: through the combination of energy recovery processing and Cu processing (reductive smelter), 98.4% of the Ag may be recovered [42]. To enable the recovery of plastics from this process, the IME devices need to be dismantled first to avoid incinerating plastics as an energy source. This separation process also brings the Ag vs plastic content in a more favorable range. IME devices from Figs.1, 4 and 7 have 0.08, 0.09 and 0.19 w-% Ag, respectively. By dismantling, the percentage of Ag in the resulting waste can be increased by a factor of ~ 7, ~ 5, ~ 13.

4.2 *Lifetime Optimization/Extension*

Circular strategies with a higher priority than recovery of materials from end-of-life products include e.g. repair, refurbishment and reuse. These focus on lifetime extension as or in a similar or lower-grade product and avoid the manufacturing of new replacement products. It should be noted that a more extensive description of the

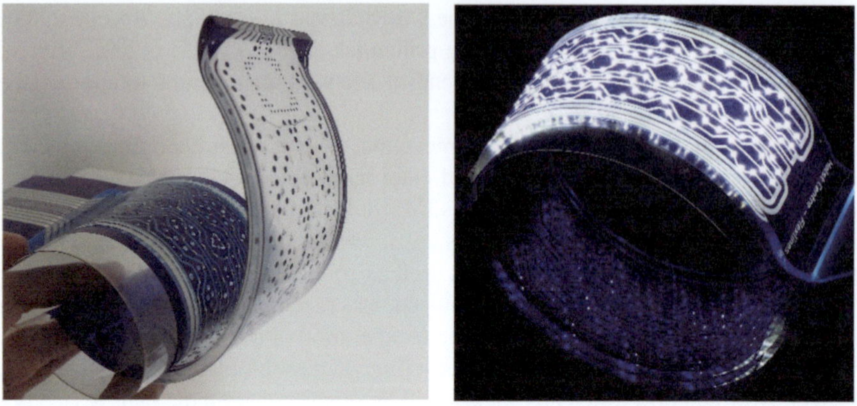

a) b)

Fig. 7 **a** Dismantled IME device with detached functional substrate and encapsulant and **b** repaired IME device

experiments and life cycle assessments are available in a recently submitted paper [19].

At the end of a product's lifetime, materials captured within need to be liberated and recovered. It is well-known that product designs may be tuned towards maximizing the recycling rates. Such design-for-recycling approaches are essential to avoid the loss of precious, strategic and critical metals [8, 34, 38, 39]. In a recent paper, we applied design-for-recycling principles to In-Mold Electronics [18] and studied the dismantling approach and potential consequences to device reliability. Lifetime extension was proposed as a logical next step. The challenge of repairing IME is like recycling: the circuitry and all semiconductor components are typically embedded within the plastics. When dismantling yields functional substrates that contain undamaged circuitry and all components in place, repair may be attempted.

In the EU project CIRC-uits, lifetime extension of IME by means of repair was investigated. Devices that were not fully functional immediately after injection molding were selected for this study. The approach to enable recycling or repair was to incorporate a dismantling layer within the device, either a water-based non-adhering dismantling layer (NADL) [18], or a water-based adhesive [19]. In the EU Unicorn project, other solvent-based commercial and research-grade adhesives were investigated. Dismantling was accomplished, either mechanically for the NADL or thermo-mechanically for the adhesive, by exposure to heat prior to applying mechanical force. The IME device was split in such a way that the functional substrate was obtained alongside the encapsulant formed by polycarbonate resin during injection molding (Fig. 7a). Glob-tops applied onto the components using the same material as the dismantling layer avoided detachment of the components from the substrate. After repair, the device could be re-encapsulated by means of injection molding (Fig. 7b).

Repairing of the functional substrates depended on the observed damages and thus varied between devices. Defects that arose from processing involved cracks in the Ag circuitry, or folds in the polycarbonate due to thermal expansion during the injection molding process: a technical issue that was resolved during the project. The microcracks stemmed from the sheer forces at high temperatures subjected to the printed layers and components during injection molding and occur close to the components. Other defects were caused by e.g. the pick and place process due to the small size of the LED components and the shaking of the LEDs in the reel during bonding with our Mycronic tool. This caused an occasional LED to be mounted upside down. The IME devices were either extensively examined and repaired (2 out of 4), minimally repaired (1 out of 4) or left untouched (1 out of 4).

Separately, a dismantled substrate with 4 large and 4 small damages to the printed Ag circuitry was repaired. This sample also needed replacing of 20 LEDs along with the conductive adhesive that was torn off along with the LEDs. These repairs were recorded in terms of time of repair and materials used, which was then used as primary data for a life cycle assessment of the repair action.

The approach to repairing the separate IME substrate was as follows: (1) dismantle the IME device in about 15 s, either mechanically or thermo-mechanically, after heating up the device for 5 min; (2) determine damages, taking roughly 60 s; (3) repair in-circuitry damage with 6.77 g of Ag ink during 3 min 50 s; (4) cure for 2 min at 120 °C in a convection oven; (5) manual ICA dispensing for 20 LEDs in 6 min 36 s using 10.93 mg Ag adhesive; (6) placement of 20 LEDs, half manually and half with the Mycronic P&P tool; (7) cure for 10 min at 120 °C; (8) manual underfilling of 20 LEDs during 6 min 20 s using 10.72 mg of epoxy adhesive in total; (9) cure for 20 min at 120 °C; (10) confirm the performance in 15 s; (11) over-mold with polycarbonate (30 s including insertion into the tool). The prescribed drying time for the material used in step (4) is 20 min, but a short drying step at this point during processing is sufficient with a large curing step at (7) and (9). LED bonding was done manually and in an automated fashion using the pick and place machine, while all could be performed with the Mycronic. Due to local warping of the polycarbonate substrate, caused by dismantling, the tool needed to be recalibrated to compensate. This required a total calibration time of 5 min and 6 s for 10 LEDs. Picking and placing of 10 LEDs after recalibration took 30 s in total. Manual bonding of the other 10 LEDs took 9 min 22 s in total. Manual repairs and automated repairs have different contributions to the environmental impact and the costs involved: manual repairs contribute heavily to the costs, but automated repairs have quantifiable contributions to power consumptions that will show up in the LCA.

The electric performance of the repaired and reference devices was measured using an oscilloscope (DSO6034A, Agilent Technologies). Vosc values for repaired devices and their references are provided in Fig. 8. Two types of reference devices are provided: without any dismantling layer (white) or with a water-based dismantling layer (cyan). For the repaired devices, two correspond to extensively repaired, one to left untouched and one to minimal repairs. The first three have a similar average Vosc, but the repaired device with minimal repairs exhibited issues related to microcracks

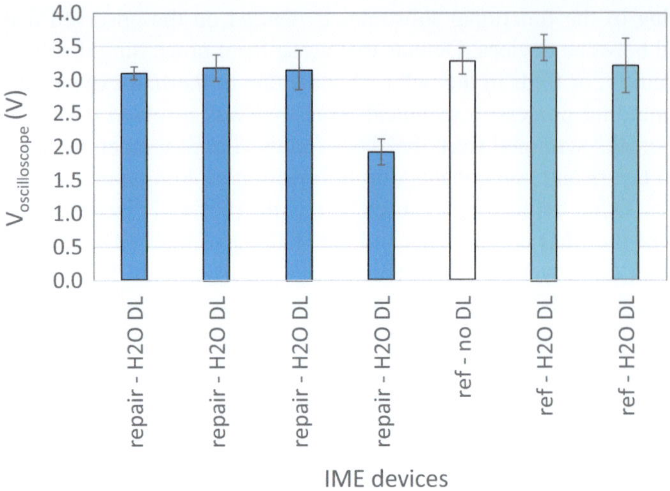

Fig. 8 Vosc for repaired IME lighting devices and references

and failing LED strings, causing the Vosc to be considerably lower. Equal electrical performance was further demonstrated with additional power measurements [19].

One question addressed in the study concerned the benefits of repair to the overall environmental impact. For this, the scenario of repairing immediately after product screening was chosen instead of repairing during the use phase. This allowed comparing two scenarios with equal life span, namely (1) a failing device that is replaced by a new device with a certain lifespan, and (2) a failing device that is repaired and has the same lifespan as the device in scenario (1). The devices in (1) would be disposed of and incinerated for recovery of heat and electricity. The devices in (2) are dismantled and repaired while the separated PC encapsulant is incinerated in scenario (2a) and recycled in scenario (2b). In scenario (2b), also the IME devices are recycled at end-of-life. With recycling, thermomechanical recycling is meant for plastics and hydro-metallurgic recycling for the Ag. Hydrometallurgy was modelled using a recent study on the LCA of EoL recycling of PCBs [22].

Figure 9 shows the relative contributions of manufacturing and EoL incineration for IME lighting devices for the recovery of heat and electricity (scenario 1). The raw materials for two devices contribute 1.72 kg CO_2 eq. to the GWP, manufacturing 0.34 kg CO_2 eq. incl IJM based on primary data for electricity and 0.21 kg CO_2 eq. of additional injection-molding impact based on secondary data (EcoInvent). End-of-life incineration provided additional burdens for both devices of 0.28 kg CO_2 kg eq. in total. Only minor benefits were obtained for a few impact categories.

For repair, especially in combination with recycling, more benefits are achievable, as shown in Fig. 10, including those stemming from the avoidance of a replacement substrate including production losses, but also from the recovery of polycarbonate and silver using thermomechanical and hydro-metallurgic recycling. Allocation of benefits from EoL in this manner is debated in literature [11] and concerns a modelling

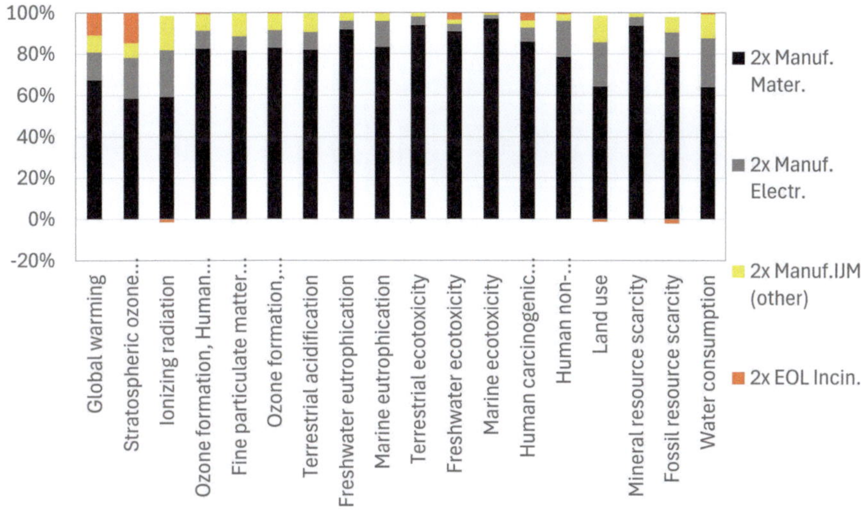

Fig. 9 Contributions to impact factors for the manufacturing and disposal of two IME devices of which one failed after manufacturing and needed to be replaced (scenario 1). At each end-of-life, the IME device is incinerated for the recovery of heat and electricity

challenge. We follow the cut-off model but extend the analysis to include a combination of circular strategies enabled by our technical solutions, thus including the benefits and burdens of repairing and recycling within the same system boundaries. Manufacturing a single IME device causes a contribution to GWP of 1.13 kg CO_2 eq. in total, while repairing an IME device contributes 0.80 kg CO_2 eq. (− 29%). Additional benefits or burdens from EoL treatment further reduce these to − 47% in case of EoL recycling and raise these to − 13% in case of EoL incineration. Additional burdens from a loss in electrical performance are relevant, as shown in Figs. 10 and 11, but do not change the overall conclusion. Experimentally, we determined that the increase in electricity consumption, relevant for the use phase, was 2 ± 2% which was low enough to avoid burdens for any of the impact categories for repairing. With a much higher increase in electricity consumption this may change, especially when incinerating obsolete parts or devices at EoL instead of recycling these.

In addition to a design that uses an adhesive or other dismantling layer, one may also use existing coatings within a device, which may be debonded using intense bursts of light. Such flashes of light may be emitted by Xenon lamps within a Pulseforge 1300 photonic curing system. The PulseForge 1300 enabled high intensity flash curing and sintering of printed metals [7], conductive adhesives and solders, but also "photonic lift-off" [2, 45]. To make repairs possible, the substrate would have to be transparent and contain a pattern of light-absorbing and light-reflecting layers. The design would have to accommodate that the circuitry would remain on the substrate, undamaged and with components still in place, while the bulk plastics are removed. Figure 12 shows an example realized on PC substrates with a flexible encapsulant (TPU/PC). Short bursts of white light were absorbed by the black graphic layer

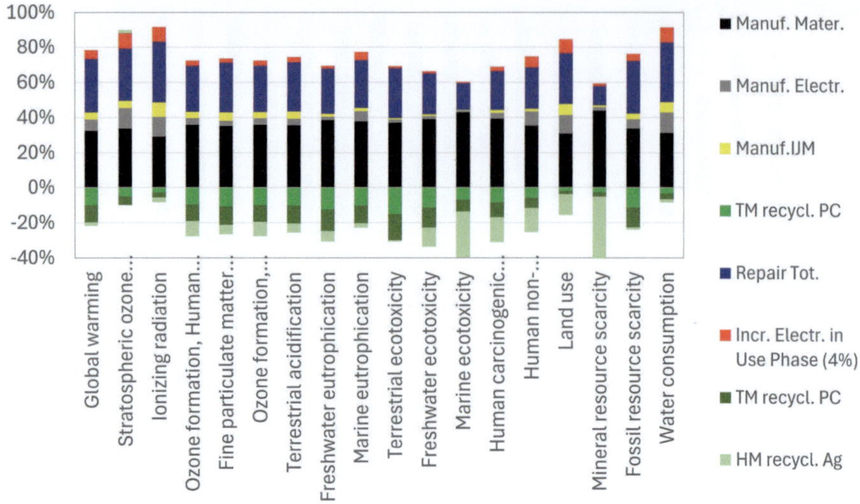

Fig. 10 Contributions to impact factors for the manufacturing and repair of an IME lighting device. At the end-of-life, the devices are recycled after dismantling (scenario 2b)

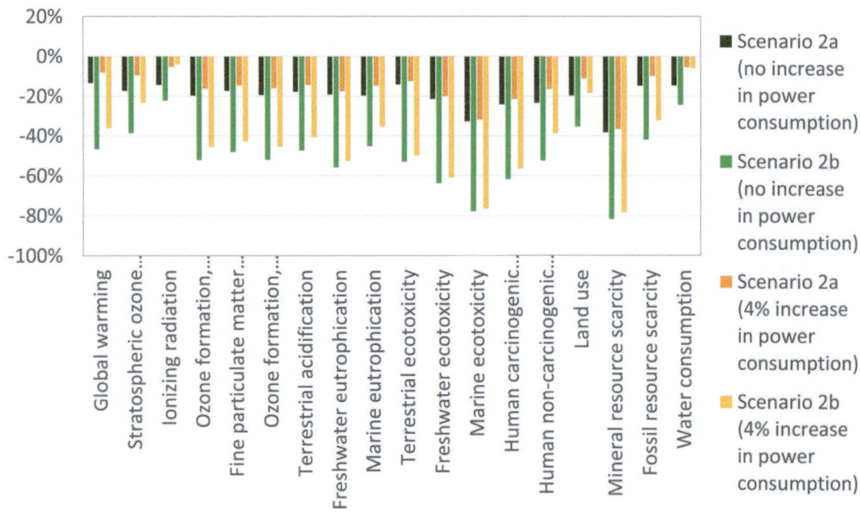

Fig. 11 Relative contributions to impact factors for the manufacturing and repair of an IME lighting device. At the end-of-life, the devices are incinerated (scenario 2a) or recycled after dismantling (scenario 2b)

Fig. 12 Delaminated lighting device achieved by photonic debonding: at the top the functional substrate with circuitry and components in place, and at the bottom the encapsulant with light-absorbing black graphic ink

and caused local delamination, while the white graphic layer underneath the silver circuitry remained unaffected. Without further effort, the encapsulant was separated from the circuitry and components. Using transparent hotmelt or adhesives within the stack, later applying one, one may be able to reattach the encapsulant after repairs. The same approach may also serve to dismantle flexible and rigid electronic devices, including IME, for improved recycling yield [20].

5 Conclusions

In this Chapter for Eco-design for In-Mold Electronics (IME), we have explored several circular pathways that contribute to a more favorable environmental impact for this hybrid & printed electronics variant. Our approach to an eco-design for IME encompasses a multitude of solutions that target decarbonization, recyclability and repairability of IME. Through the adoption of design-for-recycling principles that introduced a dedicated coating into the device design, we have been able to introduce dismantlability to an electrical device that fully encompasses the electronics within plastics. This resulted in a multitude of achievable end-of-life approaches that either recover the components, plastics and metals at end-of-life, or enable repairability, refurbishment and similar strategies. Life cycle assessments support that our approaches contribute to a reduction of greenhouse gas emissions.

Acknowledgements This research has been funded by the Dutch Ministry of Economic Affairs, and European Union's Horizon 2020 research and innovation program under grant agreements No 101003587 (EU Treasure project), No 101070169 (EU Unicorn project) and No 101091490 (EU

CIRC-uits project). The authors would like to gratefully acknowledge the contributions from Diana Godoi Bizarro, Yvonne van Liempd, Peter Rensing, Pit Teunissen, Jeroen Schram, Gerwin Kirchner and Adri van der Waal from TNO at Holst Centre. We also acknowledge the wonderful support provided by Rick Leuven and Yibo Su from Brightlands Material Center for injection molding. Finally, we gratefully acknowledge the discussions, support and recommendations related to IMSE sustainability from Pälvi Apilo and Mikko Heikkinen from TactoTek Oy.

Funded by
the European Union

References

1. Antonakou EV, Achilias DS (2013) Recent advances in polycarbonate recycling: a review of degradation methods and their mechanisms. Waste and Biomass Valorization 4:9–21. https://doi.org/10.1007/s12649-012-9159-x
2. Arutinov G, Hendriks R, van Den BJ (2016) Photonic flash soldering on flex foils for flexible electronic systems. In: Proceedings of—Electronic Components and Technology Conference 2016-August, pp 95–100. https://doi.org/10.1109/ECTC.2016.179
3. Balan A (2023) Trinseo (NL), Dissolution technology an approach for infinite circularity, advanced recycling conference 2023 Köln, pp 1–10
4. Beltrão M, Duarte FM, Viana JC, Paulo V (2022) A review on in-mold electronics technology. Polym Eng Sci 62:967–990. https://doi.org/10.1002/pen.25918
5. Brasse Y, Moreno ML, Blum S, Horter T, Janek F, Gläser K, Emmerechts C, Clanet J-M, Verhaert M, Grymonprez B, Kraus T (2024) Recyclable in-mold and printed electronics with polymer separation layers. RSC Sustain 2:1883–1894. https://doi.org/10.1039/d4su00092g
6. CENELEC (2024) A methodology to improve the recyclability rate of strategic/critical metals from car electronics
7. Cauchy-Clerc H, Gerges T, Lombard P, Semet V, Allard B, Charmeau J-Y, Cabrera M (2024) Photonic curing for in-mold electronics. In: 9th Smart plastics congress, IPC, Polymeris, Eurecat, Jan 2024, Barcelona, Spain
8. Chancerel P, Meskers CEM, Hagelüken C, Rotter VS (2009) Assessment of precious metal flows during preprocessing of waste electrical and electronic equipment. J Ind Ecol 13:791–810. https://doi.org/10.1111/j.1530-9290.2009.00171.x
9. Chandrasekaran SR, Avasarala S, Murali D, Rajagopalan N, Sharma BK (2018) Materials and energy recovery from e-waste plastics. ACS Sustain Chem Eng 6:4594–4602. https://doi.org/10.1021/acssuschemeng.7b03282
10. Ciano MP, Peron M, Panza L, Pozzi R (2025) Computers and Industrial engineering industry 4.0 technologies in support of circular economy : a 10R-based integration framework. Comput Ind Eng 201:110867. https://doi.org/10.1016/j.cie.2025.110867
11. Corona B, Shen L, Reike D, Carreón JR, Worrell E (2019) Towards sustainable development through the circular economy—a review and critical assessment on current circularity metrics resources, conservation & recycling towards sustainable development through the circular economy—a review and critical assessment on current circularity metrics. Resour Conserv Recycl 151:104498. https://doi.org/10.1016/j.resconrec.2019.104498
12. Fazlali Z, Schaubroeck D, Cauwe M, Ugduler S, Van Laere T, Manhaeghe D, De Meester S, Cardon L, Vanfleteren J (2024) Eco-friendly in-mold electronics using polylactic acid. In: 2024 Electronics goes green 2024+ (EGG), Berlin, Germany. pp 1–7
13. Fletcher CA, Dunk RM (2018) In the search for effective waste policy : alignment of UK waste strategy with the circular economy, 04:48–62

14. Gehring F, Prenzel TM, Graf R, Albrecht S (2021) Sustainability screening in the context of advanced material development for printed electronics. Mater Tech 109. https://doi.org/10.1051/mattech/2022013
15. Ghosh B, Ghosh MK, Parhi P, Mukherjee PS, Mishra BK (2015) Waste printed circuit boards recycling: an extensive assessment of current status. J Clean Prod 94:5–19. https://doi.org/10.1016/j.jclepro.2015.02.024
16. GoBizarro DE, De Simon L, Samani P, Rensing PA, Van der Waal A, Schram JFM, Van Delft JPH, Kok MM, Harkema S (2025) Life cycle assessment (LCA) of printed electronics: an in-mold electronics (IME) car touchscreen panel case study, Manuscript in progress
17. Goument C, Gerges T, Lombard P, Lakhdar H, Arli M, Martial V, Auguste S, Bruno L, Jean A, Charmeau Y, Cabrera M (2023) In-mold electronics on poly (lactic acid): towards a more sustainable mass production of plastronic devices. Int J Adv Manuf Technol 125:2643–2660. https://doi.org/10.1007/s00170-023-10878-4
18. Harkema S, Rensing PA, Domensino SMDC, Vermeijlen JM, Godoi DE, Van SA (2024) Disassembly of in-plastic embedded printed electronics. J Clean Prod 450:141837. https://doi.org/10.1016/j.jclepro.2024.141837
19. Harkema S, GoBizarro DE, Bakker MH, van Delft JPH, Ostendorf PR, Rensing PA, de Simon L, van Schaik A (2025) Repairing of in-mold electronics and life cycle assessment, submitted 11 Mar 2025
20. Harkema S, Rentrop C (2024) Recyclable electronic device, and method and apparatus for manufacturing and recycling the device, pp 1–39
21. Holst A, Lacy P, Reers J, Schmidt A, Tillemann L, Wolf C (2020) Raising ambitions : a new roadmap for the automotive circular economy
22. Iannicelli-zubiani EM, Irene M, Recanati F, Dotelli G, Puricelli S, Cristiani C (2017) Environmental impacts of a hydrometallurgical process for electronic waste treatment: a life cycle assessment case study. J Clean Prod 140:1204–1216. https://doi.org/10.1016/j.jclepro.2016.10.040
23. Ippolito M, Passadoro M, Ferella F, Pellei G, Vegli F (2023) Recovery of metals from printed circuit boards by Gold-REC 1 hydrometallurgical proces, pp 1–13
24. Ippolito NM, Passadoro M, Zueva S, Vegliò F (2024) EU treasure project D6.2: report on bio_hydrometallurgical materials recovery
25. Jääskä J, Apilo P, Otsamo K, Pirilä S, Rusanen O, Wuori T (2023) In-mold structural electronics for sustainable smart surfaces
26. Jääskä J, Muñoz I (2022) Environmental performance of IMSE. https://www.tactotek.com/resources/webinar-environmental-performance-of-imse. Accessed 8 Sep 2023
27. Kirchherr J, Reike D, Hekkert M (2017) Conceptualizing the circular economy: an analysis of 114 definitions. Resour Conserv Recycl 127:221–232. https://doi.org/10.1016/j.resconrec.2017.09.005
28. Kokare S, Asif FMA, Mårtensson G, Shoaib-ul-Hasan S, Rashid A, Roci M, Salehi N (2022) A comparative life cycle assessment of stretchable and rigid electronics: a case study of cardiac monitoring devices. Int J Environ Sci Technol 19:3087–3102. https://doi.org/10.1007/s13762-021-03388-x
29. Liu J, Yang C, Wu H, Lin Z, Zhang Z, Wang R, Li B, Kang F, Shi L, Wong CP (2014) Future paper based printed circuit boards for green electronics: fabrication and life cycle assessment. Energy Environ Sci 7:3674–3682. https://doi.org/10.1039/c4ee01995d
30. Luoma E, Välimäki M, Rokkonen T, Sääskilahti H, Ollila J, Rekilä J, Immonen K (2021) Oriented and annealed poly (lactic acid) films and their performance in flexible printed and hybrid electronics. J Plast Film Sheeting 37:429–462. https://doi.org/10.1177/8756087920988569
31. Luoma E, Välimäki M, Ollila J, Heikkinen K, Immonen K (2022) Bio-based polymeric substrates for printed hybrid electronics. Polymers (Basel) 14. https://doi.org/10.3390/polym14091863
32. Mezriczky L (2022) Trinseo: 1 month—1 member. In: European polycarbonate sheet extruders. https://www.epse.org/post/trinseo-1-month-1-member. Accessed 29 Jan 2025

33. Nassajfar MN, Deviatkin I, Leminen V, Horttanainen M (2021) Alternative materials for printed circuit board production: an environmental perspective. Sustain 13. https://doi.org/10.3390/su132112126
34. Norgren A, Carpenter A, Heath G (2020) Design for recycling principles applicable to selected clean energy technologies: crystalline-silicon photovoltaic modules, electric vehicle batteries, and wind turbine blades. J Sustain Metall 6:761–774. https://doi.org/10.1007/s40831-020-00313-3
35. Ozkan E, Elginoz N, Germirli Babuna F (2018) Life cycle assessment of a printed circuit board manufacturing plant in Turkey. Environ Sci Pollut Res 25:26801–26808. https://doi.org/10.1007/s11356-017-0280-z
36. Plast2bCleaned (2024) Plast2bCleaned project summary. https://plast2bcleaned.eu/wp-content/uploads/2024/07/Project-Sumary_Plast2_V4_compressed.pdf. Accessed 13 Mar 2025
37. Prenzel TM, Gehring F, Fuhs F, Albrecht S (2022) In fluence of design properties of printed electronics on their environmental profile, 109:1–12. https://doi.org/10.1051/mattech/2022016 (special)
38. Reuter MA, van Schaik A (2013) Resource efficient metal and material recycling. In: REWAS. Springer, pp 332–340
39. Reuter MA, van Schaik A, Ballester M (2018) Limits of the circular economy : fairphone modular design pushing the limits. In: World of metallurgy—ERZMETALL. GDMB Verlag GmbH, pp 68–79
40. Ullah M, Ippolito NM, Spera L, Vegliò F (2024) Treatment of wastewater produced during the hydrometallurgical extraction of silver from in-mold structural electronics. Case Stud Chem Environ Eng 10:6. https://doi.org/10.1016/j.cscee.2024.100916
41. van Diemen M, Brusselaers J, Rentrop C (2024) Investigating resource productivity: the ecological potential of printed electronics as alternative to PCB. Environ Res Commun 7:015035. https://doi.org/10.1088/2515-7620/ad9f0f
42. van Schaik A, Reuter MA (2023) EU treasure project D5.5: materials recycling/recovery assessment report
43. Välimäki MK, Sokka LI, Peltola HB, Ihme SS, Rokkonen TMJ, Kurkela TJ, Ollila JT, Korhonen AT, Hast JT (2020) Printed and hybrid integrated electronics using bio-based and recycled materials—increasing sustainability with greener materials and technologies. Int J Adv Manuf Technol 111:325–339. https://doi.org/10.1007/s00170-020-06029-8
44. Weeden GS, Soepriatna NH, Wang NL (2015) Method for efficient recovery of high-purity polycarbonates from electronic waste. https://doi.org/10.1021/es5055786
45. Weidling AM, Turkani VS, Akhavan V, Schroder KA, Sarah L (2022) Large-area photonic lift-off process for flexible thin-film transistors. npj Flex Electron 6:1–14. https://doi.org/10.1038/s41528-022-00145-z
46. Wiklund J, Karakoç A, Palko T, Yigitler H, Ruttik K, Jäntti R, Paltakari J (2021) A review on printed electronics: fabrication methods, inks, substrates, applications and environmental impacts. J Manuf Mater Process 5. https://doi.org/10.3390/jmmp5030089
47. Yu E, Jan K, Chen W (2023) Separation and solvent based material recycling of polycarbonate from electronic waste. https://doi.org/10.1021/acssuschemeng.3c03152

Obsolete PCB Sorting

Simonetta Cota and Viola Corbellini

Abstract The fourth pilot of the CIRC-UITS project, led by Erion, focuses on optimizing the sorting and reuse of obsolete printed circuit boards (PCBs) from Waste Electrical and Electronic Equipment (WEEE), emphasizing washing machines. Current manual sorting methods are labor-intensive and inefficient, limiting the recovery of valuable materials. This pilot introduces AI-based sorting models, a digital marketplace, and advanced human–machine interfaces to support operators and streamline the reuse process. By connecting treatment plants with manufacturers, the pilot promotes component-level reuse over traditional recycling, aligning with EU regulatory trends and contributing to the resolution of the semiconductor material crisis. Preliminary results highlight both the potential and challenges of implementing scalable circular business models. Key barriers include regulatory constraints related to end-of-waste status and the need for functional testing infrastructure. Addressing these issues is essential for mainstreaming reuse pathways and enhancing the circularity of electronics.

Keywords Obsolete PCBs · WEEE · Circular economy · Digital tools · Artificial intelligence · Component reuse · End-of-waste · Marketplace · OEMs · PCB sorting · Semiconductor crisis · Human–machine interface · Functional testing · Sustainable electronics · Secondary raw material

1 Introduction

The fourth pilot of the CIRC-UITS project focuses on obsolete printed circuit board (PCB) sorting, a critical aspect of the circular economy in the electronics sector. Currently in treatment plants, PCB can be either shredded or manually sorted. In the

S. Cota (✉) · V. Corbellini
Strategic Development and Innovation Team, ERION Compliance Organization Scarl, Milan, Italy
e-mail: simonetta.cota@erion.it

V. Corbellini
e-mail: viola.corbellini@erion.it

P. Rosa and S. Terzi (eds.), *Circular Approaches for the Ecodesign, Repair and Remanufacturing of Car and Mass Electronics*,
PoliMI SpringerBriefs, https://doi.org/10.1007/978-3-032-04301-6_8

former case, plastics and precious metals are the only materials that can be fairly recovered by processing the scraps with different methods; metal scraps are first separated magnetically from other fractions, while non-ferrous metals are extracted from mixed plastic streams using induced electromagnetic currents in a process known as eddy current separation, which exploits the repulsive forces generated in conductive metals to achieve efficient sorting. Finally, plastic fractions are separated by density to group different polymer types. In the latter case, sorting is performed manually by an operator that must be highly skilled, able to identify specific elements, like golden pins or CPU slots, on the boards to guess the value of the board and consequently, the category to which that PCB belongs. This process is extremely time-consuming. The current manual sorting procedure involves detaching gold, copper or aluminum components from the circuit board, in general components of which the operator has learned to recognize either for their value or because they are aware that specific processing streams exist, through which such "pure" material can be recovered more efficiently by following a different treatment process; the boards are then sorted based on their value into categories defined by the plant to interface with the traditional smelter market. The value of the boards is primarily determined by the amount of gold, silver, and copper visible on the board and recognizable to the naked eye. A large amount of visible gold on the board's pins indicates it is a high-value board. On the other hand, the presence of many plastic sockets suggests the absence of active components, meaning the materials present are mainly plastic and copper.

The value of the boards can also be indirectly recognized based on the sector or application field they originate from, but this it has to be considered a tailor made developed skill of the treatment plant, because is strictly related with the ability of the treatment plant to build a strategy and learn by doing process.

The treatment plant may choose to develop different processing lines; for example, it can obtain boards or parts of them by dismantling or mechanically processing electronic devices handled on the R4 line (known as cat.5 and cat.6 in Europe) (Figure 1). Alternatively, it may receive mixed boards from other plants that either do not perform clustering or only do so partially. (Figure 2) The way the treatment plant defines its categories is based on the requirements of the commercial partners it supplies. The smelter also assigns a value, and consequently a price, to each bag based on its composition, the consistency and purity of the received material, and the quality over time, reflecting the value of the contained precious metals. The pilot is designed to improve the efficiency of PCB recovery, reuse and sorting processes through advanced digital tools, artificial intelligence-based sorting models, and a dedicated marketplace. This pilot aligns with CIRC-UITS' broader goal of addressing the semiconductor material crisis by enhancing value chains in the automotive and mass electronics industries.

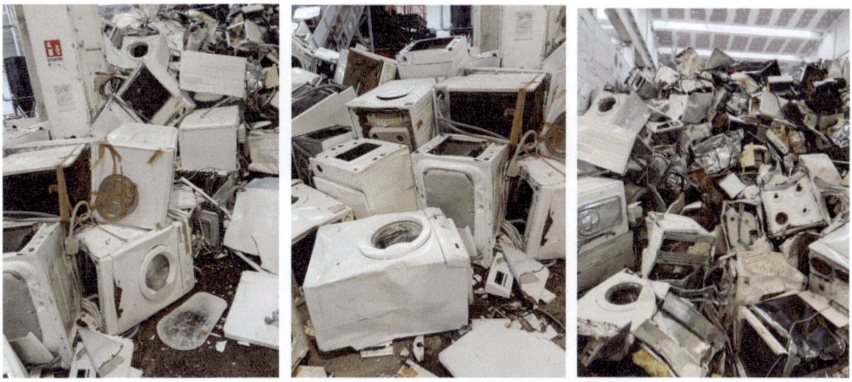

Fig. 1 Washing machine condition at the first treatment plant involved in the scouting of PCBs

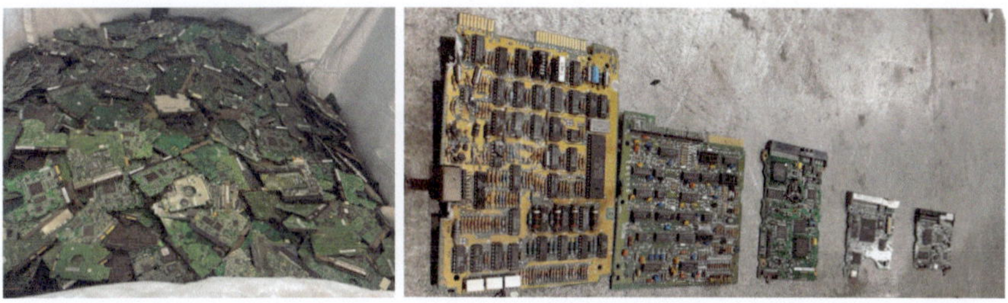

Fig. 2 Left: batch of HDD PCBs received by an external supplier to be clustered by the treatment plant according with the different value. Right: HDD PCB boards divided by year (from left to right, oldest to newest) reflecting the evolution of manufacturing technologies over time and the increasing capability to integrate microelectronics, thereby reducing component volume and the amount of material used

2 The Importance of Obsolete PCB Sorting in CIRC-UITS

The pilot plays a crucial role in showcasing how circular business models can be integrated into the electronics sector, particularly within WEEE treatment plants. Erion, the largest PRO system for WEEE in Italy, leads the fourth pilot of the project, titled "Obsolete PCB Sorting."

As shown in Figure 3, the pilot focuses on assessing the reusability of electronic circuit boards and their components recovered from washing machines. At the same time, a dedicated marketplace is being developed to track the availability of PCBs suitable for reuse, creating a direct connection between treatment facilities and electrical and electronic equipment (EEE) manufacturers.

The pilot also tests improvements to the current selection process for electronic boards sourced from mixed WEEE streams intended for recycling. To support this, the project develops digital solutions that assist operators during the selection process, while also evaluating the operational and economic benefits of these innovations.

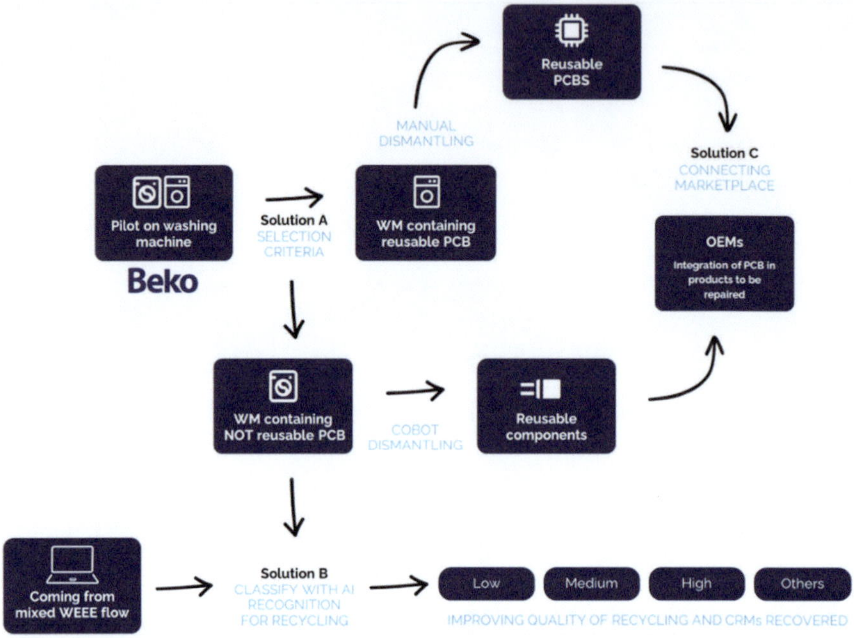

Fig. 3 Pilot 4 scheme

The use of artificial intelligence aims to improve the sorting process by increasing both the speed and consistency in assembling batches of PCBs destined for smelter facilities. Additionally, AI reduces reliance on operator expertise, making the process more efficient and replicable. The core idea behind training the AI to recognize specific features of PCBs is also leveraged to assess the feasibility of sourcing components from the end-of-life (EoL) sector. This supports the pilot's goal of enabling PCB reuse in manufacturing by OEMs, promoting a shift from traditional recycling to component-level reuse in line with evolving EU regulations. A major challenge lies in the lack of structured pathways for Original Equipment Manufacturers (OEMs) to source reusable PCBs. Currently, PCBs in large household appliances (LHAs) are typically extracted after the initial shredding of the machine. While this process facilitates the extraction of components such as the motor, concrete counterweight, and the PCB itself, it does not guarantee that the extracted PCBs remain undamaged. This method of processing materials has never truly addressed the challenge of preserving product integrity beyond the so-called "use phase." As a result, it has led not only to the loss of valuable materials, critical to the digital and green transitions, on which global manufacturing relies; the whole system is not designed to consider whether recovering functional components, and potentially repairing them, could be more economically viable than traditional material recovery methods. For this reason, within the CIRC-UITS project, Pilot 4 is committed to recognize the value hidden in electronic waste, exploring the untapped potential of this stream. The pilot aims to investigate a business model in which a marketplace platform is developed to bridge this gap by creating a centralized digital environment where treatment plants and OEMs can interact, facilitating the reuse of extracted components like PCBs.

This digital platform enables the identification and exchange of reusable PCBs, streamlining the supply chain for reclaimed components. The marketplace relies on AI-driven image recognition for PCB classification and integrates with an HMI system to facilitate real-time identification and processing. The tool is designed to promote circular economy practices by enhancing the availability of refurbished components in the short term while ultimately reducing reliance on new semiconductor materials in the long term.

3 Assessment and Future Implications

The pilot establishes assessment criteria for sorting and reusing PCBs, considering parameters such as material composition, functional testing, and historical production data. Initial testing phases have demonstrated challenges in extracting targeted PCBs due to variations in waste stream conditions. However, the findings highlight the potential for optimizing sourcing strategies and improving digital tools for future scalability. As EU regulations demand higher reuse and recycling targets, the results from this pilot will inform policies and operational strategies, ensuring that circular business models become standard practice in the electronics industry. In this context, several barriers still need to be overcome one of the most critical is how to manage the materials leaving a treatment facility, particularly whether they can be classified as “end-of-waste” and therefore made eligible for reintroduction into a marketplace system. Given the current structure of the end-of-waste framework, the only viable way to activate a business model like that of Pilot 4 would be for the manufacturer to subcontract a treatment facility to operate as a test lab. In this setup, products, still not classified as waste, would be sent to the facility through a dedicated private logistics system for testing and potential reuse. While a take-back scheme could technically be developed, this approach risks “privatizing” the concept of large-scale repair, potentially undermining broader goals related to accessibility, transparency, and circularity. Meanwhile, the preparation-for-reuse pathway enables facilities to carry out these operations and officially declare end-of-waste status for tested and repaired components. While this could represent a viable solution, the volume of components sufficient to justify the investment required to systematize a new process in the treatment plant remains uncertain. Moreover, this route is currently more bureaucratically complex, requiring additional regulatory adaptations to ensure a streamlined and effective implementation.

PCB Remanufacturing Activities

Paolo Rosa, Lorenzo Gandini, and Sergio Terzi

Abstract In this chapter, we will present the activities carried out by POLIMI in parallel with various pilot plants to automate the remanufacturing process of certain electronic components and in addition inspection of circuitry. Specifically, the activities related to the testing materials provided by partners of the CIRC-UITS project will be shown, including ECU from Bosch, the in-tire sensor from Continental, the washing machine board from BEKO, and the in-mold electronics from TNO. The basic idea is to explore automated solutions using Key Enabling Technologies such as robotics and machine learning to enable the automation of the task of repairing/remanufacture electronic components from PCB boards. The study aims to shed light on whether it is possible to automate a task currently performed only by highly skilled labor and to highlight the potential limitations or challenges in developing such a solution. In the initial phase of activities, feedback from manufacturers will be gathered to define the disassembly targets for the products under examination. Once the targets are received, we will proceed with selecting the appropriate tools for disassembly, choosing from the precision rework tools for electronics available in the Polimi's Industry 4.0 Laboratory and considering the development of custom solutions specifically designed for the target components. Manual disassembly tests will then be conducted to evaluate the performance of the selected tools, identify any issues, and define the correct strategies for disassembly. After completing the manual disassembly of the components, they will be sent back to the partners, who will test them to evaluate their functionality. By creating a closed-loop process around disassembly, continuous feedback and bilateral communication will enable the definition of optimal procedures to automate subsequently. In parallel with the disassembly activities, solutions based on machine vision will be studied to identify the desired components. This will allow for greater flexibility in the automated solution, as these algorithms will guide the disassembly process and eliminate the need for reconfiguration for each different layout of PCBs that may contain the same target component.

P. Rosa · L. Gandini (✉) · S. Terzi
Department of Management, Economics and Industrial Engineering, Milan, Italy
e-mail: lorenzofrancesco.gandini@polimi.it

P. Rosa and S. Terzi (eds.), *Circular Approaches for the Ecodesign, Repair and Remanufacturing of Car and Mass Electronics*,
PoliMI SpringerBriefs, https://doi.org/10.1007/978-3-032-04301-6_9

Keywords Remanufacturing, AOI, AI, Robotics, PCBs

1 Current State of the Art on Electronics Remanufacturing

The burgeoning issue of electronic waste (e-waste) presents substantial environmental and economic challenges, which have spurred research into effective recovery and reuse strategies. As a critical technological solution, automated disassembly of electronic components is emerging to enhance the efficiency of recycling processes. This brief literature review critically examines recent advancements in the field, with a focus on automated processes that facilitate the reuse and remanufacturing of electronic components. Stobbe et al. [1] offer a foundational perspective on the necessity for quality assurance in automated disassembly systems. They advocate for a model in which quality-assured disassembly supports the reuse of electronic components, underscoring the necessity of integrating robust quality control mechanisms to ensure the viability of recovered components for reuse. This approach not only promotes sustainability but also enhances economic feasibility by diminishing the costs associated with manual disassembly processes emphasizing the critical importance of the ability to verify the operational status of components following disassembly actions. Advancing the discourse, Saenz et al. [2] identify specific requirements essential for the effective automated disassembly of e-waste. Their research underscores the importance of integrating detailed modeling of disassembly processes and product states, which is crucial for developing systems capable of adapting to diverse electronic products and components. Their findings suggest that intricate process modeling is pivotal for optimizing disassembly lines to manage the complexity and diversity of modern electronic waste. Considering this, it becomes evident that technologies capable of abstracting and operating on different products are of fundamental importance. Specifically, the use of artificial intelligence-based solutions could assist in accomplishing this complex task. Shahbazi et al. [3] explores practical applications through a case study on electric and electronic equipment in Sweden, investigating how product design significantly affects the efficiency of automated remanufacturing processes. They conclude that principles of design for disassembly (DfD) should be integral to product development to enable easier, more cost-effective disassembly and recovery of valuable components. The effectiveness of end-of-life electronics treatment is strongly influenced by the design phase, it becomes clear that a closed-loop approach is necessary to connect the end-of-life stage with the beginning-of-life stage. This perspective aligns with the broader theme that automation technologies must be supported by appropriate product designs to maximize their effectiveness. Zhao et al. [4] critically review various techniques for disassembling and reusing components from printed circuit boards (PCBs, their evaluation of manual, semi-automatic, and fully automated disassembly methods concludes that while automation promises enhanced efficiency, it also demands significant advancements to manage the intricate and densely packed nature of modern PCBs. Modern PCB boards are, in fact, highly complex components, composed of many elements situated on very small surfaces. This complexity significantly complicates the selective disassembly of components without affecting others on the board. This issue is particularly critical when attempting to repair a PCB, where the goal is to replace the damaged component while preserving the integrity of the entire board. Finally, Marconi [5] examine the feasibility of designing an automated disassembly system. Their findings suggest that while automated systems can significantly improve the disassembly process, they must be tailored to specific types of electronic components and devices to optimize recovery rates and minimize waste. Consistent with the previously stated observations, it becomes clear that automated systems used for the removal of components from the board must be specifically designed for the component in question. The current difficulty in creating a flexible solution imposes significant limitations on the development of an optimal solution, necessitating dedicated studies when aiming to handle a wide range of components. This limitation suggests that, at present, it is more practical to focus only on the most expensive and systemically important components. The literature collectively indicates a growing consensus on the potential of automated disassembly systems to revolutionize e-waste management. However, significant challenges remain in terms of technology development, system integration, and economic viability. Future research should focus on refining these technologies, enhancing their adaptability to various e-waste categories, and improving their economic attractiveness to encourage widespread adoption. This review reflects a nuanced understanding of the complex interplay between technology, process optimization, and economic factors in the context of electronic waste management, highlighting both the progress and the hurdles that remain to be overcome.

2 Methodology

Precision robotics play a pivotal role in ensuring the accurate separation of delicate components, streamlining the overall remanufacturing workflow. The research moves beyond traditional manual testing methods by introducing closed-loop feedback mechanisms, which not only enhance flexibility and precision but also reduce the dependency on skilled labor. This shift towards automation establishes a more sustainable and economically viable remanufacturing process. Ultimately, the integration of these advanced technologies is expected to revolutionize POLIMI's remanufacturing operations, setting a new standard for future endeavors in the electronic manufacturing sector. The methodologies developed throughout this research are poised to significantly improve the precision, efficiency, and sustainability of remanufacturing practices, aligning with the broader goals of Industry 4.0. For this activity, it was necessary to create a specific model that would allow closed-loop feedback to be created with the manufacturers of the boards being tested. This was necessary since testing activities performed on disassembled components can only be carried out by manufacturers who have the hardware and especially software resources to verify the status of the components (Fig. 1).

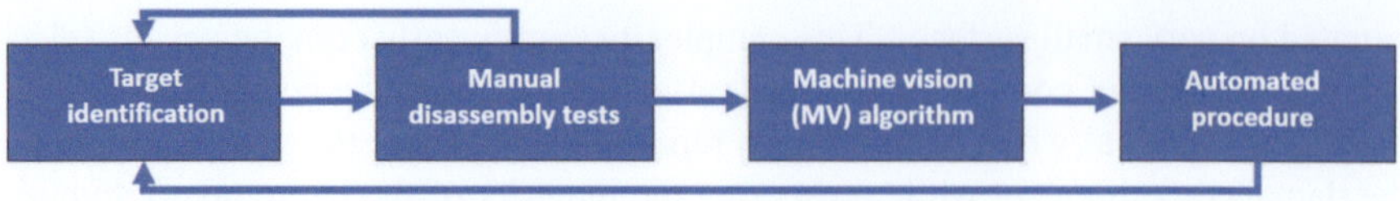

Fig. 1 The workflow adopted for the remanufacturing activities

The workflow commences with a comparative analysis alongside original equipment manufacturers (OEMs). This step is essential for identifying the target components on which remanufacturing activities will be performed. It necessitates collaboration with experts possessing in-depth knowledge of the product, both from functional and economic perspectives, to guide the remanufacturing process effectively. During this phase, critical information is collected regarding the components' assembly techniques and relevant disassembly guidelines. These guidelines establish the operational parameters, such as permissible temperature limits and temperature ramp profiles, which must be adhered to ensure safe and effective handling of the components (Fig. 2).

Once preliminary information is collected, manual disassembly tests are conducted. These tests allow for a preliminary feasibility study aimed at identifying potential critical disassembly issues in the target components. For this purpose, different precision tools are used, and different methodologies are tried to remove the component as efficiently as possible. it is at this point that the first feedback is

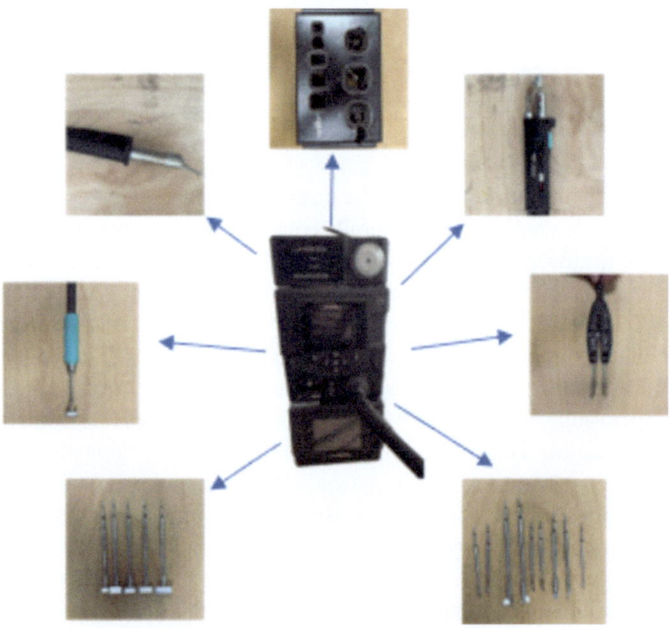

Fig. 2 The precision rework tools used during manual disassembly tests

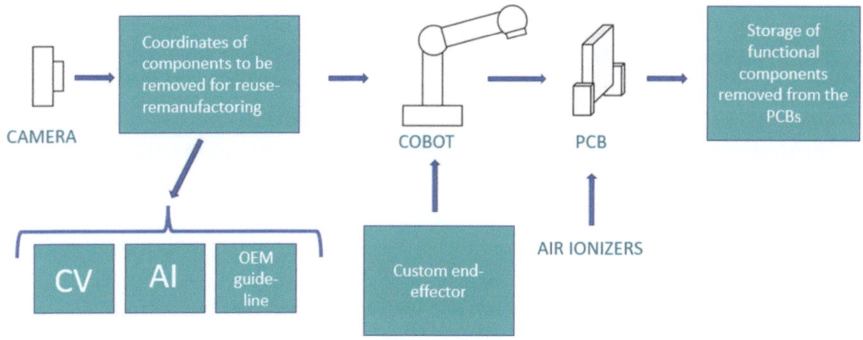

Fig. 3 Proposed framework for the pilot plant dedicated to remanufacturing activities

collected from the manufacturers by sending the disassembled material for functionality testing. Consequently, the use of AI-based solutions for the identification of target components is assessed. This phase is considered optional, as the necessity for development depends on the specific type of component targeted for recovery. If the component is found only in a particular PCB layout, the added value of an autonomous identification system is minimal. Conversely, if the component is generic and present in various PCB layouts, it may be worthwhile to invest in the development of an automated identification system. However, it is essential to verify the availability of the data required to train these models, as such data is often inaccessible or lacks the necessary level of detail to support the process effectively (Fig. 3).

The final phase involves the development of a proof of concept for an automated system aimed at component removal. The core aspect of this activity is the prototyping of end effectors for the removal of specific components. This task is carried out in collaboration with external organizations to develop custom solutions, which are produced using additive manufacturing techniques. This process will be carried out iteratively, testing the solution and gathering feedback from manufacturers to refine and optimize its functionality. A thorough analysis of the technologies is also essential to determine the appropriate type of equipment required for the system.

3 Use Cases

3.1 Bosch ECU

Upon receiving the material from Bosch, consisting of PCB boards from ECUs extracted from their plastic cases, we gathered the manufacturer's inputs regarding the target components for removal and disassembly procedures. The received ECU boards come in four different layouts and feature various types of electronic components. The components chosen as targets for removal are the integrated circuits on the ECUs, as they are the most valuable components on the board. Additionally, to simplify preliminary operations, we decided to focus exclusively on the larger components. The defined objective is to remove the components while preserving the integrity of both the components and the source board.

For soldering and desoldering specifications, we adhered to the JEDEC J-STD-020D.1 standard, which imposes limits on the maximum temperatures that components can reach, and the temperature ramp limits to avoid potential thermal shock. Bosch also recommended using air ionizers during disassembly activities to prevent possible damage to the boards from static electricity (Fig. 4).

Once alignment with the manufacturer was completed, we began initial manual disassembly tests to better understand which tools would be suitable for the remanufacturing task of these boards. These tests are crucial for defining the actuators that will perform the component removal in automated disassembly. The performance of the tools was evaluated based on disassembly speed, stress on the components and boards (assessed by the temperatures reached by both during disassembly), and the visual appearance of the parts after the procedures were carried out. The target components, although all belonging to the category of integrated circuits, come in two different configurations: Ball Grid Array (BGA) and Quad Flat Pack (QFP). Analyzing the board layout, it was decided to proceed with disassembly using the hot air desoldering tool. The tests, conducted on all presented layouts, were performed

Fig. 4 Target components proposed by Bosch

in two different manners depending on the type of component; the QFPs were disassembled by directing the hot air stream directly onto the component until it was disassembled, while the BGAs, due to their composition, were heated from the backside of the board to allow the heat to diffuse through the component from below. Both methodologies present trade-offs: direct hot air is faster in disassembling the component but exposes the component to higher temperatures, risking damage, whereas rear-applied hot air takes more time and desolders components on the backside of the board as well but allows for a more uniform temperature distribution within the component. The tests were always performed with thermocouples placed on the components to monitor temperature trends. Following these tests, the removed components and boards were sent to Bosch for functionality testing. Overall, this layout proved particularly complex to disassemble due to the high density of components on the board. This issue sometimes prevented the use of heat sinks to direct the heat properly and led to the unintentional disassembly of non-target components located near the heat application point. The difficulty in positioning the heat sinks also translated into challenges in placing the component extractors. This resulted, particularly in the case of BGAs, in the solder spreading over the component surface during the removal procedure, necessitating cleaning of the component before it could be retested.

In parallel with the disassembly activities, a computer vision-based solution was developed to identify the target components. Since the defined targets are all integrated circuits, a network was designed to recognize and locate them on the PCB. The first step was to identify a dataset containing annotations of integrated circuits on PCBs to provide the network with training data application; the WPCB-EFA dataset Pramerdorfer and Kampel [6] was identified as a candidate dataset, this dataset contains data on ICs in image and coordinates format. "The dataset contains 748 images of PCBs from a recycling facility, captured under representative conditions using a professional DSLR camera. For all these images we provide accurate segmentation information for the depicted PCBs as well as bounding box information for all Integrated Circuit (IC) chips (9313 samples)". Following an analysis of models in the literature and aware of the complexity of using Convolutional Neural Networks (CNNs) for recognizing electronic components, it was chosen to use a procedure like the one illustrated by Silva et al. [7]. We chose to use the YOLOv5 model, performing transfer learning on the network's final layers to adapt it to our specific needs. Using models like YOLO, which are pre-trained on large datasets, transfer learning proves particularly effective. By training only, the final layers of the network, we can achieve high performance even with a relatively small amount of provided data. This approach leaves the initial layers of the network, responsible for recognizing basic features (points, lines, contours, and basic shapes), unchanged while focusing on higher-level classification tasks (Fig. 5).

After training the model, its performance was evaluated, and tests were conducted on the boards to assess its effectiveness. The model performed well, identifying all target components except one specific component (Fig. 6).

This component, being a next-generation integrated circuit with a significantly different visual appearance from the circuits annotated in the training dataset, could

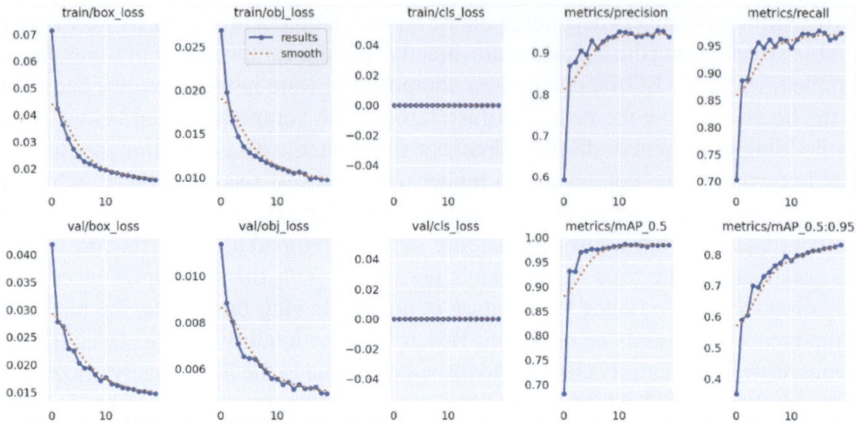

Fig. 5 Performances and metrics of the network reported in graphical form highlights the results of the transfer learning performed

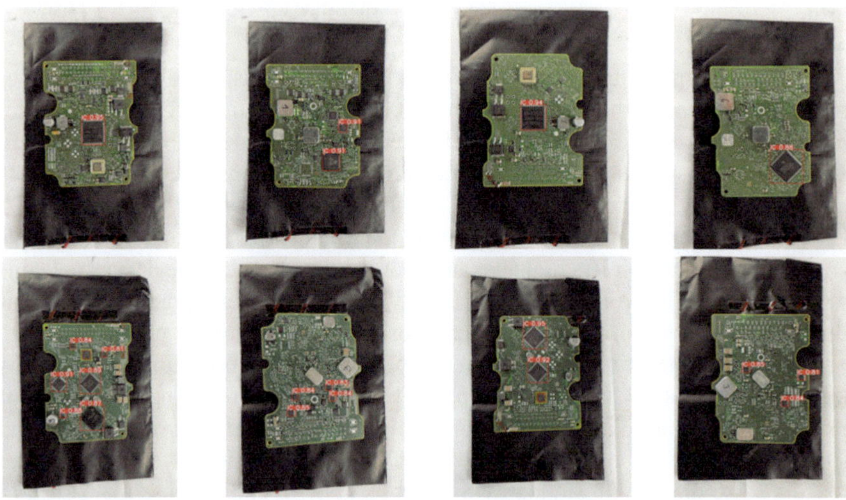

Fig. 6 In red the bounding boxes containing the target components (integrated circuits)

not be consistently identified by the model. To address this issue, an alternative approach for identifying this component was studied.

This approach, based on conventional computer vision techniques, involved searching for specific features of the component in the image. Fortunately, the component has an orange frame, a color relatively easy to isolate from the rest of the board. An algorithm was developed to track this color and thus denote the component's position. The use of these two solutions enabled the realization of the first component of the automated solution: the identification of components necessary to guide the system to the target positions. Additionally, using AI-based algorithms provides the flexibility needed to operate on different board layouts while always targeting integrated circuits.

3.2 Continental In-Tire Sensor

The remanufacturing project for the Continental In-Tire sensor was conducted to explore potential rework and reuse of components within automotive sensor technology. The focus was on determining the feasibility of disassembling and reworking specific sensor components, particularly the Printed Circuit Board (PCB), to support sustainability and cost-effectiveness within the automotive industry. A complete disassembly of the Continental In-Tire sensor was conducted, providing a preliminary mass balance that outlined the component breakdown. The sensor initially weighed a total of 1.46 g, with post-disassembly results indicating that the PCB had retained 0.96 g of the total mass. This process also identified that smaller components posed significant challenges during disassembly, which was critical in assessing the potential for remanufacturing.

The challenges posed by smaller components impacted the feasibility of disassembly. The delicate and intricate nature of the PCB required precise handling, indicating that only larger components might be practical for rework or repair. Manual disassembly methods were employed to define the best procedures for potential automation. Manual methods were particularly important due to the small size of the board and the complexity involved in detaching components without damage. Specialized equipment for precision rework applications was leveraged to facilitate the removal of candidate components (Fig. 7).

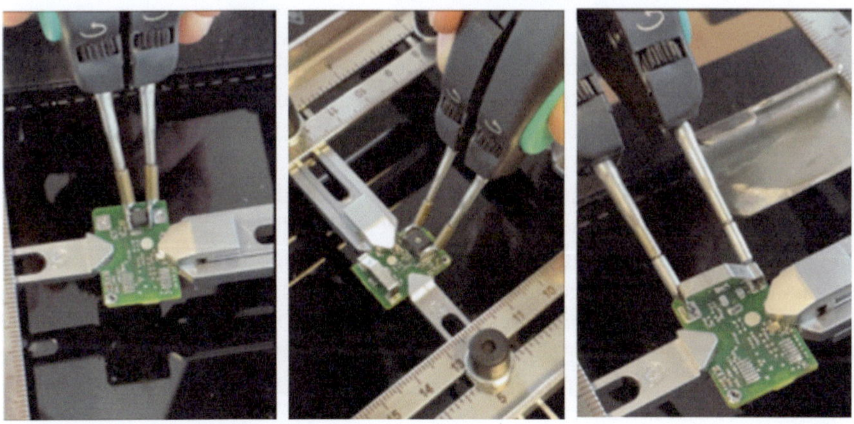

Fig. 7 Disassembly of biggest components of the PCB

However, despite extensive testing, the remanufacturing process for this sensor proved to be highly complex. The primary challenges identified were:

- Component Size: The very small size of certain components made it difficult to manage without damaging other parts.
- Complexity of Automation: Automated remanufacturing tasks were rendered impractical due to the intricacy and sensitivity required for disassembly.
- Suspension of Remanufacturing Activity: In coordination with Continental, it was determined that the project's remanufacturing efforts would be suspended, as they were deemed infeasible under current conditions and resources.

The project's findings underscore the challenges in remanufacturing complex automotive sensors such as the Continental In-Tire sensor. While larger components like the PCB are theoretically viable for rework, the technical complexity and practical limitations of disassembling smaller parts pose substantial barriers. Future research and development could explore the following:

- Enhanced Automation Tools: Developing specialized tools and machinery that can handle small components more effectively.
- Component-Level Optimization: Designing sensors with remanufacturing in mind, such as modular components that can be easily disassembled.
- Material Recycling Alternatives: If remanufacturing is not viable, investigating effective recycling methods for sensor materials may provide a more feasible solution for sustainability.

The suspension of this activity highlights the need for more advanced technologies to support sustainable practices in automotive sensor remanufacturing.

3.3 *BEKO Washing Machine PCB*

The remanufacturing project for BEKO washing machine Printed Circuit Boards (PCBs) was designed to explore effective strategies for component recovery and reuse. This analysis is critical in the context of sustainable practices and reducing electronic waste within the home appliance industry. The primary focus was on disassembling Surface-Mounted Device (SMD) components, particularly targeting components labeled QD2 and QD5, as these were identified as significant candidates for potential reuse and remanufacturing. Initially, the scope of target components was broad, which posed challenges in streamlining the disassembly process. Refining the selection to individual components such as QD2 and QD5 allowed for more focused tests and procedures. By narrowing the scope to these specific SMD components, the project aimed to maximize efficiency and improve the chances of successful recovery (Fig. 8).

Several disassembly tests were conducted using different types of equipment. Key equipment utilized included:

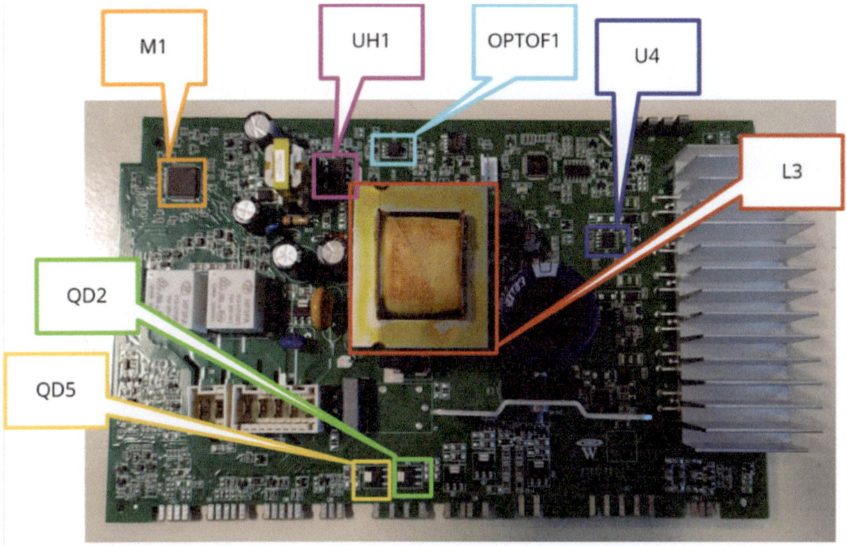

Fig. 8 The washing machine PCBs and the initial target components

- Custom IC Desolderer: This tool was essential for precision removal of integrated circuits from the PCB.
- Air + Heat Sink Combination: This method was tested to assess its efficiency in mitigating thermal damage during disassembly (Fig. 9).

The disassembly process revealed significant issues, particularly during the removal phase. Components often exhibited residual solder paste, especially around the pins. This necessitated the use of a specialized component extractor to achieve cleaner removal, albeit at the expense of increased complexity for automation. The residual solder pastes on components post-removal indicated a need for more

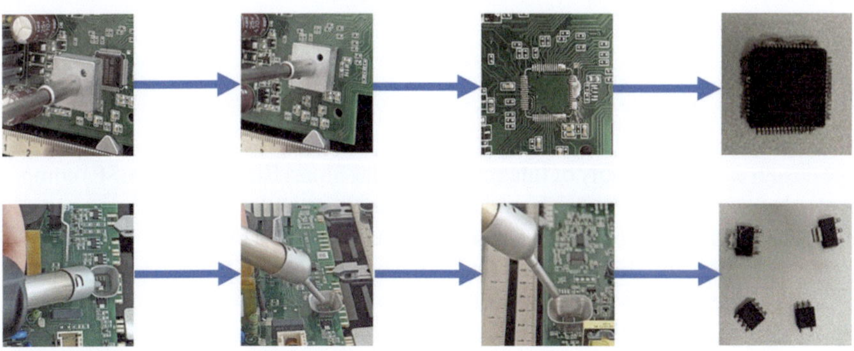

Fig. 9 Results of the two preliminary disassembly methods

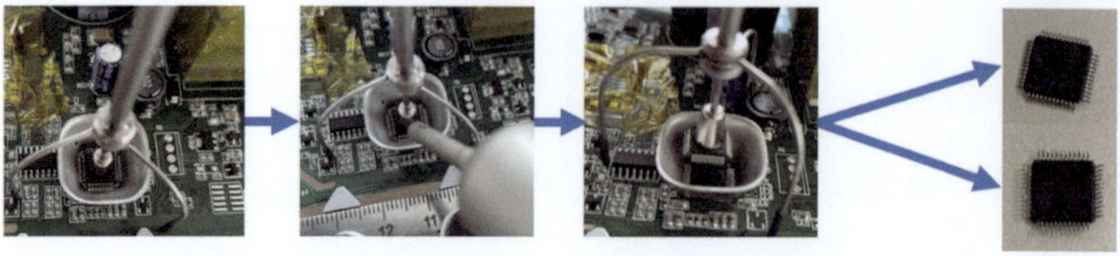

Fig. 10 The new disassembly procedure leveraging component extractor for clean removal of ICs

advanced extraction techniques. Integrating a component extractor improved the quality of removal and provided greater operational efficiency (Fig. 10).

However, this addition increased the complexity of automating the disassembly process. Despite this challenge, the plan is to test a custom end-effector previously developed for the target components of Bosch ECU to evaluate its effectiveness in automating the extraction of components. The BEKO washing machine PCB remanufacturing project underscored several key points:

- Complexity of SMD Removal: Manual disassembly proved viable but highlighted difficulties related to residual solder paste. Automating the process, while possible, would require sophisticated equipment to manage these issues.
- Component-Level Testing: Testing specific end-effectors for automated disassembly could streamline the process, making it feasible for larger-scale applications.
- Sustainability Implications: Successfully remanufacturing washing machine PCBs can contribute to reduced electronic waste, aligning with broader sustainability goals.

Future recommendations include exploring advanced desoldering techniques and refining automation tools to handle component-specific challenges. By continuing to optimize these processes, BEKO could further enhance its remanufacturing capabilities, leading to improved sustainability and resource efficiency in appliance production.

3.4 IMSE Inspection

The design and disassembly of Integrated Modular Smart Electronics (IMSE) components play a critical role in modern manufacturing, particularly in enhancing sustainability through remanufacturing and recycling efforts. This section outlines the quality control methodologies explored in the context of IMSE disassembly, emphasizing the potential of visual inspection technologies to improve disassembly accuracy and efficiency. The activities focus on creating a structured approach to disassemble IMSE components while ensuring that quality control is rigorously maintained throughout the process. A key objective is to identify faults or defects that may occur after disassembly, which can affect the remanufacturing potential of these components (Fig. 11).

The core of the proposed approach involves the investigation of visual inspection technologies that can detect surface-level defects or the absence of critical components. One of the technologies being explored is one of the most advanced smart camera systems available on the market. These cameras are capable of high-precision surface defect detection, identifying the presence or absence of components with considerable accuracy.

In collaboration with the smart camera provider, feasibility studies have been conducted to explore the capabilities of this smart camera technology. The primary focus of these studies was on detecting missing LEDs, a common defect during the disassembly process. The results of these studies were promising, demonstrating that multiple areas of inspection could be defined to ensure thorough quality control. The successful identification of missing LEDs marked a key milestone in validating

Fig. 11 Catalog of defects provided by TNO

the feasibility of this technology for use in IMSE disassembly. However, this study represents only the initial step in a broader quality control strategy. Future work will expand beyond detecting missing LEDs, targeting additional defects such as surface scratches and errors that occur during the pick-and-place operations in automated assembly systems and solve any potential misalignment of the sample from its nominal shape.

The smart camera system utilized in this research represents cutting-edge technology in visual inspection. These cameras are designed to work seamlessly in automated environments, leveraging advanced algorithms to detect minute defects that might otherwise be overlooked by manual inspection processes. The technology's capacity to adapt to varying inspection needs makes it an ideal tool for ensuring the quality of disassembled IMSE components. Multiple inspection zones can be set up for each component, allowing for real-time analysis and immediate detection of potential faults. This capability is particularly useful in complex manufacturing environments where rapid feedback is crucial for maintaining production quality and avoiding delays (Fig. 12).

While the results of the feasibility study are encouraging, several challenges remain in the implementation of this technology on a broader scale. These include:

- Surface Defect Detection: The detection of scratches and other surface imperfections will require further refinement of the algorithms used by the Keyence

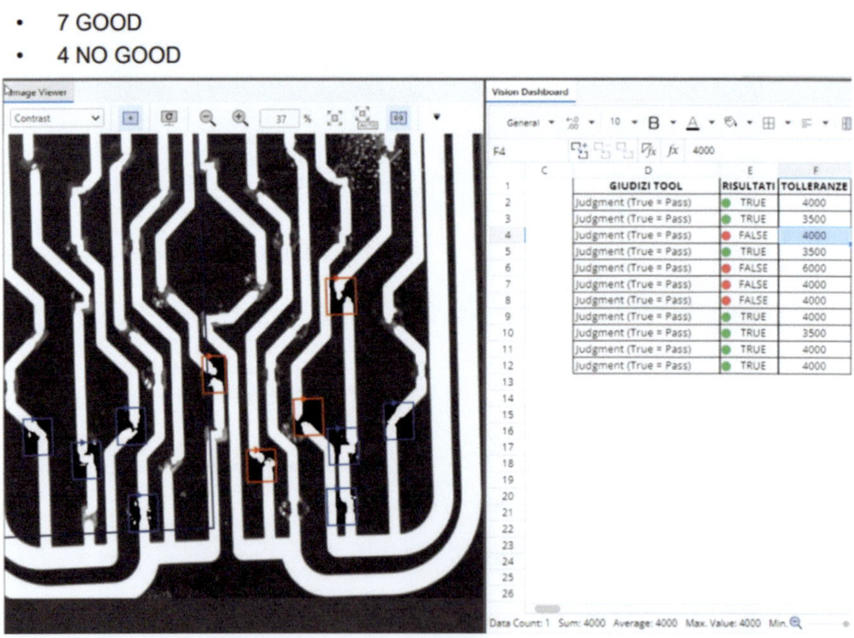

Fig. 12 Outcome of the missing components assessment. The red (missing component) and blue (component) squares on the left image are the areas of inspection

cameras. These defects are often more subtle than missing components and thus harder to identify accurately.

- Pick-and-Place Errors: The integration of quality control into automated pick-and-place systems is another area requiring further investigation due to possible deformation of the board because of thermal cycling. Identifying errors during this stage is crucial for ensuring the proper assembly of remanufactured components with pick and place machines.
- Scalability: As with any advanced technology, scaling the use of smart cameras for defect detection across a full production line presents both technical and economic challenges. Further studies will be needed to assess the cost–benefit ratio of widespread implementation. The exploration of visual inspection technologies represents a significant advancement in quality control for the disassembly and remanufacturing of IMSE components. The ability to accurately identify missing components, such as LEDs, highlights the potential of smart camera systems to enhance the disassembly process. Future work will focus on refining these techniques to detect more subtle defects, ensuring the long-term viability of this approach. By improving quality control during disassembly, the aim is to contribute to more sustainable manufacturing practices, particularly in the context of the growing need for remanufacturable and recyclable electronic components.

4 Future Developments

The future direction of this research will focus on advancing automation techniques and developing new technologies for more effective electronic component remanufacturing. Specifically, the following areas will be prioritized:

- Prototyping Robotic End-Effectors for Component Removal: A major development effort will focus on designing and prototyping robotic end-effectors specifically for removing the Bosch ECU and BEKO PCB target components. These end-effectors will be tailored to handle different component configurations such as BGA and QFP, incorporating feedback from the manual testing phase to optimize their effectiveness.
- Testing of Automated Procedures: Once the robotic end-effectors are prototyped, extensive testing of the automated removal procedures will be conducted. This will involve both real-time trials on PCB components as well as simulation environments to fine-tune control parameters. The objective is to ensure that the automated processes achieve efficiency levels comparable to or exceeding those of skilled labour, while maintaining component integrity.
- Total Automation of IMSE Repair: Efforts will also be directed toward achieving full automation in the repair of IMSE components. Specifically, this will involve resolving the sample deformation problems identified during manual disassembly activities. Techniques to assess deformation, To be able to communicate possible changes in the position of components to the pick and place machine that will re-assemble them.

5 Conclusions

The activities presented in this chapter underscore the significant potential for automating the disassembly and remanufacturing of electronic components, but also highlight the substantial challenges that must be overcome to make these processes commercially viable. The manual disassembly trials serve as a critical foundation for understanding the complexities of each component and layout, which directly inform the development of specialized tools and automation strategies. Our case studies, particularly with the Bosch ECU and BEKO PCB, demonstrate the advancements made in using precision robotics and computer vision-based solutions to guide disassembly processes. However, these studies also expose the limitations that persist in areas such as high component density, residual solder paste, and variability in component design.

This research contributes to the growing body of work on electronic component remanufacturing by providing detailed insights into the practical challenges of automating processes that are traditionally reliant on skilled labor. Specifically, the findings demonstrate that while automation technologies—such as robotic end-effectors and machine vision systems—offer significant promise, the variability and complexity of modern electronic devices present substantial barriers to full automation. The study also reinforces the importance of product design for disassembly (DfD), supporting existing literature that advocates for the integration of remanufacturing considerations at the early design stages to maximize recovery potential. Furthermore, the development and testing of AI-based solutions for component identification contribute to the field by showcasing how artificial intelligence can increase the flexibility and efficiency of remanufacturing processes, although the need for comprehensive, high-quality data remains a critical challenge.

The implications of this research extend beyond the immediate technical challenges of automation to broader environmental and industrial contexts. As e-waste continues to grow at an alarming rate, the development of efficient, scalable remanufacturing processes is essential for reducing landfill waste and mitigating the environmental impact of electronic products. By focusing on the automation of disassembly and component recovery, this study provides a pathway toward more sustainable management of electronic waste. These advancements align with the goals of Industry 4.0, particularly in terms of creating smart, interconnected production systems that can adapt to complex tasks such as remanufacturing.

The integration of robotics, machine learning, and computer vision in remanufacturing workflows is a clear step towards the future of manufacturing, where automated processes not only enhance efficiency but also contribute to circular economy principles. However, the findings also highlight the need for continued research and innovation to address the technological and economic challenges that currently limit widespread adoption. Future efforts should focus on the refinement of robotic tools, the improvement of AI-driven component identification systems, and the development of flexible automation solutions capable of handling a wide variety of electronic components.

In conclusion, while significant progress has been made in automating the disassembly and remanufacturing of electronic components, substantial hurdles remain. Addressing these challenges will require interdisciplinary collaboration and ongoing innovation in both technology and product design. The continued evolution of these systems will be critical for enabling the large-scale adoption of automated remanufacturing processes, ultimately contributing to more sustainable electronic waste management and aligning with the long-term objectives of Industry 4.0.

References

1. Stobbe I, Griese H, Potter H, Reichl H (2002) Quality assured automated disassembly of electronic components for reuse. IEEE
2. Saenz J, Felsch T, Walter C, König T (2024) Automated disassembly of e-waste—requirements on modelling of processes and product states. Front Robot AI
3. Shahbazi S, Johansen K, Sundin E (2021) Product design for automated remanufacturing—a case study of electric and electronic equipment in Sweden. Sustainability 13(16):9039
4. Zhao W, Xu J, Fei W, Liu Z, He W, Li G (2023) The reuse of electronic components from waste printed circuit boards: a critical review. Environ Sci: Adv
5. Marconi M, Palmieri G, Corinaldi D (2019) Feasibility study and design of an automatic system for electronic components disassembly. J Manuf Sci Eng 141(2):021011
6. Pramerdorfer C, Kampel M (2015) A dataset for computer-vision-based PCB analysis. In: 2015 14th IAPR international conference on machine vision applications (MVA), Tokyo, Japan, pp 378–381. https://doi.org/10.1109/MVA.2015.7153209
7. Silva L, Júnior A, Azevedo G, Oliveira S, Fernandes B (2021) Estimating recycling return of integrated circuits using machine vision on PCBs. Appl Sci 11:2808. https://doi.org/10.3390/app11062808
8. Kirillov A, Mintun E, Ravi N, Mao H, Rolland C, Gustafson L, Xiao T, Whitehead S, Berg AC, Lo W-Y, Dollár P, Girshick R (2023) Segment anything

Circular Business Models Identification and Development

Laura Pomo, Daniele Perossa, Laila El Warraqi, and Paolo Rosa

Abstract In the CIRC-UITS project, Circular Business Models (CBMs) have been developed to facilitate the transition towards circular economy (CE) in the electronics and automotive sectors. A CBM is defined as a model that creates, captures, and delivers value by improving resource efficiency through extending the life of products and closing material loops. The business models developed during the European Project CIRC-UITS promote the innovation applied to automotive and electronics industries, particularly semiconductors, within new value chains. Through the implementation of digital tools such as data-sharing platforms, blockchain for traceability, and digital twins, companies can monitor the lifecycle of materials and improve resource efficiency. An investigation was conducted on companies and European projects engaged in the remanufacturing of electronic components across two main sectors: electric vehicles (EV) and mass electronics. This research was divided into three steps: the first one covered the identification of relevant companies and projects in sectors aligned with the project goals, the second one focused on the collection of the business models information using the Business Model Canvas (BMC) framework based on available public data. Lastly, the study summarized key strategies currently adopted or potentially implementable by these companies to enhance sustainability practices. The proposed circular business models encourage greater collaboration among companies and support the creation of markets for secondary raw materials, while also reducing dependence on virgin resources and non-EU supplies. Additionally, these models integrate lifecycle analysis, environmental impact assessments and product design to guide businesses toward more sustainable practices.

Keywords Circular business models · Circular economy · Circular business model framework · Reverse supply chain

L. Pomo (✉) · D. Perossa · L. El Warraqi · P. Rosa
Department of Management, Economics and Industrial Engineering, Politecnico di Milano, Milan, Italy
e-mail: laura.pomo@polimi.it

P. Rosa and S. Terzi (eds.), *Circular Approaches for the Ecodesign, Repair and Remanufacturing of Car and Mass Electronics*,
PoliMI SpringerBriefs, https://doi.org/10.1007/978-3-032-04301-6_10

1 Introduction

In recent years, the concept of CE has gained significant attention from policymakers and businesses, positioning itself as a framework that balances economic growth with environmental sustainability [5]. CE emphasizes the minimization of resource consumption and waste generation by extending product lifecycles and closing material loops [6]. To implement CE on an organizational level, CBMs play a central role. They thrust companies to rethink their value propositions and design value chains that prioritize cost efficiency, production effectiveness, and overall sustainability [7]. Among the various CBM typologies, Product-Service Systems (PSS) are particularly promising for their ability to shift the focus from product ownership to service provision, enabling companies to retain control over product lifecycles and encourage sustainable usage [15]. PSS models are generally classified into three main categories: product-oriented services, where customers purchase a product with complementary maintenance services, use-oriented services, which involve leasing or sharing arrangements; and result-oriented services, where users pay for the outcome or functionality provided by a product rather than for the product itself [14]. Within the framework of the CIRC-UITS project, CBMs—especially PSS-based and 3R (Reuse, Remanufacturing, and Recycling) models—are being developed to facilitate the electronics and automotive sectors' transition toward CE. The goal of this study is to explore and promote these CBMs, evaluating their potential to reduce environmental impact, optimize resource utilization, and enhance business resilience in line with CE principles [11].

2 Current State of the Art on CBMs

The increasing demand for green solutions in modern industries has placed circularity and sustainability at the forefront of strategic business development, particularly in sectors such as electronics and automotive. CBM, which incorporate strategies like product repair, refurbishment, and closed-loop supply chains, are becoming crucial for creating value while adhering to sustainability principles [2].

This chapter presents an in-depth analysis of current business models in these sectors, drawing from a comprehensive review of academic literature, European projects for business use cases, and company practices. It explores the methods used to develop new business models that prioritize sustainability while also delivering economic value.

The research presented here forms a crucial foundation for understanding how companies can reimagine their business models to align with circular economy principles, ultimately contributing to both environmental and economic sustainability. This work is part of the European project CIRC-UITS, which aims to promote sustainable development goals through innovative industry practices.

2.1 Research Methodology

The research methodology employed in this study is divided into two main steps: an extensive review of academic and industrial literature; an empirical analysis of real-world business practices in the industrial world and in European projects.

Literature Review: The literature review focused on 25 key papers that examined circularity, sustainability, and business modelling within the mass electronics and automotive industries. These papers provided a theoretical base, highlighting the importance of integrating sustainable and circular practices into traditional business models [3].

Industry Analysis: The research also incorporated an analysis of 18 European projects and 48 company websites across relevant sectors. This analysis was conducted to understand how businesses currently approach circularity and to identify gaps that could be addressed through new circular business models.

Together, these research components provided a comprehensive view of the current business models state of the art.

The Business Model Canvas (BMC), Osterwalder and Pigneur [8] has been used as a strategic tool to visualize available technical information by addressing key components such as value propositions, customer relationships, revenue streams, and key resources of the companies or projects studied. As part of this research, the BMC framework was applied to analyze both European projects and companies operating in the electronics and automotive sectors.

European Projects: The analysis yielded five tailored BMCs, each designed to assess the business strategies implemented in specific European projects. These canvases provided insights into how these projects aim to integrate circularity and sustainability into their business models.

Companies: Additional six BMCs were developed for companies operating within sectors relevant to the CIRC-UITS project. These models highlighted current business strategies and identified areas where sustainability initiatives could be strengthened.

For what concerns European Projects, the analysis produced five customized BMCs, each designed to assess specific European projects focused on circularity within the electronics and automotive industries. Notable projects include Sustainably SMART, which emphasizes modular design for smartphone recycling; DiCiM (Mass Electronics), which offers a tiered pay-per-use model for washing machines incorporating predictive maintenance; and SCANDERE and DiCiM Automotive, which develop data platforms for tracking financial incentives to support the return and reuse of used automotive parts. Additional projects, such as CE-RISE, UNICORN, AUTO TWIN, DaCapo, CircThread, Circular TwAIn, and Sustronics, contribute to sustainability by promoting resource recovery and circularity across various technological fields. Each BMC provided insights into how these projects integrate sustainable practices by focusing on key partnerships, innovative revenue models, and value propositions that align with the goals of the European project CIRC-UITS. Regarding Companies, six BMCs were created for companies in electronics and automotive sectors, allowing an in-depth look at the circularity strategies they implement. Companies such as Alec GmbH, My Auto Solutions and E-repair showcase advanced remanufacturing processes, emphasizing transparent, automated workflows and rigorous quality control. Focus is placed on R&D to consistently improve processes, with an emphasis on innovation and achieving circularity goals. AES Modules provides specialized automotive repairs with a strong environmental focus, while ISS (name) and Actronics highlight innovations in logistics, reverse supply chains, and simulation technologies that optimize component durability and performance. The analysis clearly shows that for all companies studied, innovation is highly relevant, with technological advancements frequently acting as primary drivers of sustainable solutions [9]. Further analysis of these companies revealed unique approaches to circularity, from easy regeneration services to consultancy investments in R&D for diagnostics and maintenance enhancements. This comprehensive BMC application to both European projects and industry players offers a clear view, based on the current business realities and aiming to enhance their effectiveness from a circular perspective, of how circular economy principles can be effectively integrated into business models, supporting operational efficiency, regulatory compliance, and long-term environmental sustainability within the CIRC-UITS framework.

3 Business Models Development and Results Obtained

The CE concept has been extensively explored, with Geissdoerfer et al. [4], defining CBM as systems designed to cycle, extend, intensify, and dematerialize material and energy flows. These principles aim to reduce resource inputs while minimizing waste and emissions, thereby aligning business operations with sustainable development goals. Moreover, according to [5], by adopting sustainable business models, companies can enhance their financial, social, and environmental performance, thereby increasing their resilience and reducing their exposure to external risks.

Building on these concepts, a comprehensive review and comparison of academic literature on business model innovation frameworks has been conducted. The framework developed by [1], was selected to represent the business models developed within the CIRC-UITS project due to its thorough integration of sustainability and circularity principles, which align closely with CIRC-UITS' core objectives.

The framework is structured across three levels: the business ecosystem level, the business level, and sustainability impact, as represented in Fig. 1 [1].

The Business Ecosystem Level emphasizes an analysis of external trends and drivers that support circularity. Key elements include market demands, technological advancements, and regulatory changes, all of which contribute to shaping the business landscape. Additionally, the stakeholders' environment is critically examined,

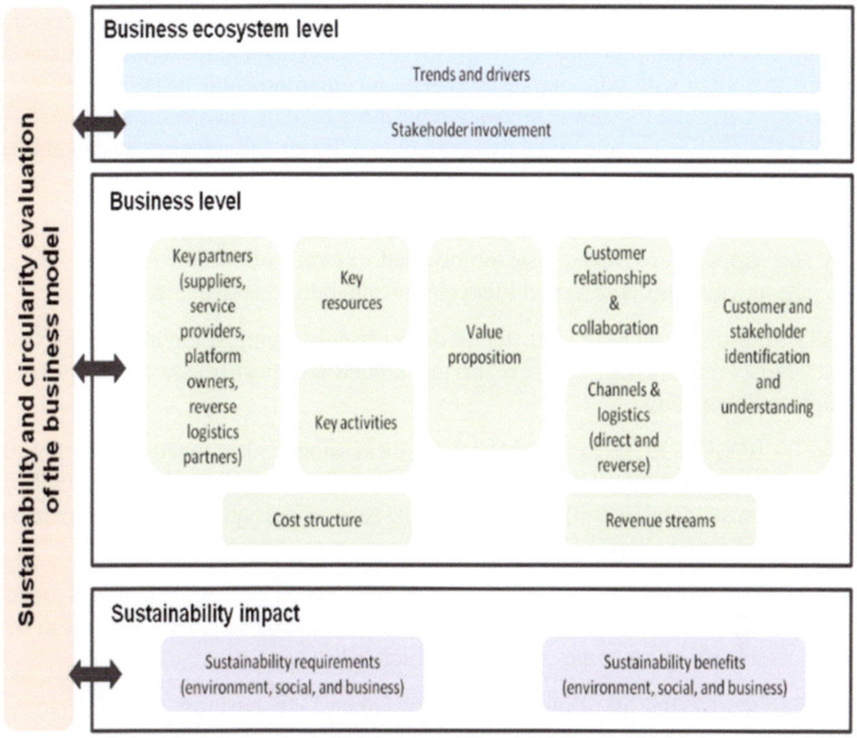

Fig. 1 Business model framework for sustainability and circularity evaluation

engaging external entities such as governments and policymakers, whose regulations and legislation significantly influence sustainability practices.

The Business Level is the house of the core components of the traditional Business Model Canvas (BMC), Osterwalder and Pigneur [8], are retained, but with an expanded integration of the stakeholders' perspective. This expansion recognizes that stakeholders, including suppliers, partners, and the broader community, have a significant influence on business activities, extending beyond traditional customer relationships.

The Sustainability Impact focuses on sustainability requirements and sustainability benefits. Sustainability requirements involve the collection and analysis of data on sustainability metrics, such as lifecycle assessments (LCAs), to evaluate the environmental, social, and economic impacts of business practices. The sustainability benefits are derived from the implementation of circular practices, including reductions in resource consumption, the extension of product lifecycles, and decreases in CO2 emissions.

These features align closely with the primary goals of CIRC-UITS, which seeks to develop innovative business models that promote resource efficiency, waste reduction, and the creation of closed-loop systems. In this chapter, we will explore the results that this framework application brought in the context of the project, demonstrating how its principles have been applied to develop sustainable, circular business models that align with both market demands and environmental sustainability.

Within CIRC-UITS, seven innovative business models have been developed to address the challenges of critical raw materials (CRMs) and advance sustainability and circularity practices across various sectors. These models include:

Digital Toolbox business model: A platform providing functionalities related to advisory and support for CRMs, disassembly, and recovery of components, as well as lifecycle assessments (LCA) and lifecycle sustainability assessments (LCSA).

OFFIS Spin-off business model: focused on offsetting supply, environmental, and social impact risks through advisory services related to criticality assessment, product design, and recycling.

Pilot 1—BOSCH ECU Repair Model business model: developed together with Bosch, this model emphasizes the repair and reuse of electronic control units (ECUs), enhancing sustainability and cost-effectiveness for original equipment manufacturers (OEMs) and repair shops.

Pilot 2—Tire Information System business model: owned by Continental, this model employs a product-as-a-service (PaaS) approach, ensuring sensors remain with the manufacturer for reuse at the end of the vehicle's life.

Pilot 3—Eco-Design and Manufacturing of Green IME business model: Tracxon utilizes a product-service system (PSS) to facilitate the repair and recycling of in-mold electronics (IMEs), fostering sustainability in production.

Pilot 4—Obsolete PCBs Sorting business model: BEKO focuses on enhancing the sustainability of washing machines through the reuse of printed circuit boards (PCBs), thereby reducing reliance on virgin materials.

Serious Games business models: these models aim to educate users about the transition to a circular economy, fostering awareness and interest in sustainable practices.

In the following sections, the insights gained from these models will be synthetized, highlighting their impact on sustainability and the broader implications for organizations aiming to adopt circular economy principles.

3.1 Results

The development of circular business models is crucial to align economic growth with the concept of sustainability, especially in industries such as mass electronics and automotive [10], where waste generation and resource consumption are significant. The European Project CIRC-UITS introduces innovative Key Exploitable Results (KERs) to address these challenges, each leveraging circular economy principles to extend product lifecycles, optimize resource use, and mitigate the environmental impact of production and end-of-life processing. The effectiveness of the strategies implemented to promote circularity will determine the extent to which these goals are achievable [10].

This chapter presents business models developed for these KERs, illustrating how they support the transition to a circular and sustainable economy by highlighting the innovation and sustainability advantages they offer, as well as their key business elements.

4 Digital Toolbox's Business Model

The Digital Toolbox developed by TXT represents a pivotal resource for companies aiming to integrate circularity into their business practices. The toolbox is composed of modules, developed thanks to the knowledge and the expertise of SUPSI (Scuola universitaria professionale della Svizzera italiana), Offis—Institute for Information Technology and MARAS (Material Recycling and Sustainability) B.V., and consolidates multiple functionalities that support circular economy assessments, including lifecycle analysis, critical raw materials (CRM) risk management, recycling assessment, Design for Recycling and component disassembly guidance. The value proposition of the Digital Toolbox lies in its capacity to streamline sustainability practices, combining different disciplines and LCS&CA methodologies and KPIs and regulatory compliance, thereby enhancing users' environmental, social, and governance (ESG) performance.

From the business model development work emerged:

- Innovation and Sustainability Advantages: The toolbox contributes significantly to improvement in sustainability by reducing the environmental footprint of companies and ensuring regulatory compliance with EU standards. By supporting data and knowledge driven decision-making, it enhances transparency and optimizes resource usage, resource recovery and circularity (re-use, repair, remanufacturing) across the supply chain, while at the same time revealing the limits to CE as dictated by nature laws. Additionally, it provides an accessible way for users to understand and implement sustainable practices, making it a powerful tool for advancing circular economy goals.
- Key Business elements: Revenue is generated through sale of results and platform or specific bundles (SaaS logic) and consulting services, while continuous software updates and user support ensure high adaptability for evolving industry standards.

5 Digital Twin for CRM Management's Business Model

The CRM management tool developed by OFFIS, utilizing digital twin technology, provides critical material insights for manufacturers, helping them design sustainable products with optimized CRM usage.

From the business model development work emerged:

- Innovation and Sustainability Advantages: This tool addresses the growing need to mitigate CRM shortages by promoting sustainable design decisions and early life-cycle interventions. By using digital twins, companies can analyze resource needs and assess the end-of-life impact of materials, offering a model that emphasizes both economic sustainability and resource conservation.
- Key Business elements: Revenue streams include licensing fees, sale of software (SaaS), and consulting services. Partnerships with software developers and environmental experts expand its functionality and reach. Continuous feedback will drive improvements, enabling updates and expansions tailored to customer needs. To engage clients, a demo version will be available, inviting exploration and early interaction with the tool's capabilities.

6 BOSCH ECU Repair Model's Business Model

BOSCH's business model for repairing electronic control units (ECUs) focuses on refurbishing these components and reintegrating them or their recoverable components into vehicles, extending their lifecycle.

From the business model development work emerged:

- Innovation and Sustainability Advantages: This model provides a sustainable alternative to ECU replacement by refurbishing components instead of discarding them, reducing waste, and conserving critical materials. This approach minimizes environmental impact and aligns with circular economy goals through resource efficiency.
- Key Business elements: Revenue is generated from payments by OEMs and repair services, which represent the company's primary customers.

7 Tire Information System's Business Model

This business model focuses on developing a sustainable tire information system by enabling sensors to be reused, thus extending the lifecycle of these electronic components. The system provides critical data on tire conditions, supporting timely maintenance decisions while minimizing electronic waste.

From the business model development work emerged:

- Innovation and Sustainability Advantages: The model is designed to conserve resources by enabling the reuse and maintenance of tire sensors. By facilitating sensor reuse, this system reduces demand for new materials and limits electronic waste, aligning with circular economy objectives in automotive technology.
- Key Business elements: Revenue streams include the sale of durable sensors, subscription services for tire condition monitoring data, and strategic partnerships with automotive manufacturers to integrate these solutions into their vehicles. A digital twin of tire sensors will be instrumental in optimizing eco-design and enhancing disassembly processes.

8 Sustainable IME Production's Business Model

TRACXON's model focuses on developing recyclable and repairable In-Mold Electronics (IME) for the automotive industry, offering a circular solution for electronic component manufacturing.

From the business model development work emerged:

- Innovation and Sustainability Advantages: TRACXON's IMEs are designed for easy repair and recycling, significantly lowering the use of raw materials and supporting component recovery. This eco-friendly approach reduces energy consumption and offers a sustainable design option for electronics in automotive applications.
- Key Business elements: Revenue is generated through the sale of eco-designed IMEs and maintenance partnerships with Original Equipment Manufacturers (OEMs) that prioritize sustainability. By integrating recyclable components, manufacturers can achieve cost savings while also benefiting the environment. Additionally, specialized knowledge and equipment are essential for conducting repairability assessments of the IMEs prior to their proper repair. This process utilizes advanced digital technologies, including AI tools, to effectively identify damages on the IMEs.

9 Obsolete PCBs Sorting System's Business Model

Developed in collaboration with Erion and owned by BEKO, this model aims to demonstrate the feasibility of sourcing spare parts from the end-of-life (EoL) sector. Through the use of digital sorting technology capable of detecting specific PCB characteristics, the model seeks to reintroduce recovered printed circuit boards (PCBs) for reuse in washing machines.

From the business model development work emerged:

- Innovation and Sustainability Advantages: By sorting and reusing PCBs, this model prevents valuable materials from becoming waste and promotes sustainable

use of resources. It plays a crucial role in extending the lifecycle of household appliances and significantly reducing e-waste generation enabling the exchange of components by connecting end-of-life operators with manufacturers, through the digital marketplace.
- Key Business elements: Revenues include selling Valuable PCBs as spare parts to be reintroduced in refurbished washing machines. The AI-powered sorting systems add value by ensuring that only high-quality, reusable PCBs re-enter the supply chain and PCBs that are not suitable for direct reuse can still be efficiently further enhancing resource recovery.

10 Serious Games' Business Models

The Serious Games initiative is a training model designed to raise awareness and educate students and professionals on the circular economy and sustainable business practices.

From the business model development work emerged:

- Innovation and Sustainability Advantages: These games play a crucial role in spreading circular economy knowledge, preparing future professionals to implement sustainable practices within their industries. The games simulate real-world scenarios, fostering hands-on learning and skill development in sustainability.
- Key Business Elements: The model's revenue streams include selling potentially valuable PCBs to manufacturers for testing and refurbishment, selling recovered PCBs as spare parts for maintenance operators, selling refurbished washing machines, incorporating tested and repaired PCBs as spare parts.

In conclusion, the development of innovative circular business models, as shown by the CIRC-UITS task related to business model developments, underscores the vital role of sustainability and circularity in addressing the challenges posed by high-waste industries like mass electronics and automotive [13]. These models demonstrate practical applications of circular economy principles and emphasize the necessity of technological advancements and collaborative efforts to optimize resource use and extend product lifecycles. By using tools such as the Digital Toolbox, businesses can enhance their environmental, social, and governance (ESG) performance while navigating the complexities of regulatory compliance. Moreover, the emphasis on education and training through initiatives like Serious Games highlights the importance of preparing future professionals to implement sustainable practices. Overall, these efforts pave the way for a more resilient and sustainable economy, aligning economic growth with environmental stewardship.

11 Conclusion

Developing innovative, circular, and sustainable business models is essential for driving the transition toward an economy that is not only more sustainable but also forward-looking. The seven circular business models detailed in this chapter, derived from the CIRC-UITS project, provide practical solutions for various sectors by integrating circular principles and addressing critical sustainability challenges. These models, supported by the "Framework for Sustainable Circular Business Model Innovation," showcase how companies can optimize resource use, reduce environmental impact, and create new value streams. The ongoing collaboration among CIRC-UITS partners ensures that these business models are both innovative and aligned with the latest industry practices, contributing significantly to the achievement of global sustainability goals.

References

1. Antikainen M, Valkokari K (2016) A framework for sustainable circular business model innovation. Technol Innov Manage Rev 5(7):1–8. https://doi.org/10.22215/timreview/1000
2. Florian L-F, Gold S, Bocken NMP (2018) A review and typology of circular economy business model patterns. J Ind Ecol
3. Geissdoerfer M, Vladimirova D, Evans S (2018) Sustainable business model innovation: a review. J Clean Prod 198:401–416. Available at: https://doi.org/10.1016/j.jclepro.2018.06.240
4. Geissdoerfer M, Morioka SN, de Carvalho MM, Evans S (2018a) Business models and supply chains for the circular economy. J Clean Prod 190:712e721. http://linkinghub.elsevier.com/retrieve/pii/S0959652618311867
5. Geissdoerfer M et al (2020) Circular business models: a review. J Clean Prod 277:123741. Available at: https://doi.org/10.1016/j.jclepro.2020.123741
6. Lieder M, Rashid A (2016). Towards circular economy implementation: a comprehensive review in context of manufacturing industry. J Clean Prod 115:36–51. Available at: https://doi.org/10.1016/j.jclepro.2015.12.042
7. Nußholz JLK (2017) Circular business models: defining a concept and framing an emerging research field. Environ Sustain Appl Sustain 9:1810. https://doi.org/10.3390/su910181
8. Osterwalder A, Pigneur Y (2010) Business model generation. Wiley. https://doi.org/10.1017/CBO9781107415324.004
9. Pollard J et al (2021) A circular economy business model innovation process for the electrical and electronic equipment sector. J Clean Prod. https://doi.org/10.1016/j.jclepro.2021.127211
10. Prochatzki G et al (2023) A critical review of the current state of circular economy in the automotive sector. J Clean Prod. https://doi.org/10.1016/j.jclepro.2023.138787
11. Rosa P, Sassanelli C, Terzi S (2019) Circular business models versus circular benefits: an assessment in the waste from electrical and electronic equipments sector. J Clean Prod 231:940–952. https://doi.org/10.1016/j.jclepro.2019.05.310
12. Rosa P, Terzi S (2021) New business models for the reuse of secondary resources from WEEEs. The FENIX project. SpringerBriefs in Applied Sciences and Technology. http://www.springer.com/series/11159
13. Suppipat S, Hu AH (2022) Achieving sustainable industrial ecosystems by design: a study of the ICT and electronics industry in Taiwan. J Clean Prod 369(March):133393. Available at: https://doi.org/10.1016/j.jclepro.2022.133393

14. Tukker A (2004) Eight types of product-service system: eight ways to sustainability? Experiences from Suspronet. Bus Strat Environ 13:246–260
15. Tukker A (2015) Product services for a resource-efficient and circular economy—a review. J Clean Prod. https://doi.org/10.1016/j.jclepro.2013.11.049

Standardization Actions in CIRC-UITS

Sarah Köhler and Christian Grunewald

Abstract The following chapter gives an overview about standardization in general, its processes and products. It also describes how standardization can support innovation and can be used in research projects, like CIRC-UITS. As an excerpt from the standardization landscape for CIRC-UITS, existing standards, ongoing standardization projects, as well as relevant standardization committees (technical committees) are presented in the context of the automotive and the electronics sector under consideration of circular economy and environmental aspects. The needs and gaps in standardization identified by the CIRC-UITS project are listed and ideas for filling these gaps are explained. The need for standardization in relation to activities regarding the digital product passport (DPP), the repairability and the assessment of repairability of products, as well as the need for standardization considering circular economy approaches in the field of rare earth were identified. Standardization can support the dissemination and exploitation of project results and thus the implementation of project innovations on the market through standardization activities, e.g. by contributing to ongoing standardization projects at national, European or international level or by developing new standardization documents. Various standardization activities have been performed in order to close different gaps in the standardization field. Project results dealing with a DPP of printed circuit boards and the aspect of the improvement/assessment of a product repairability, are planned to transfer to standardization through the development of two standardization documents. First standardization activities have been initiated. The description of planned and conducted standardization activities within CIRC-UITS completes the chapter on standardization.

S. Köhler (✉) · C. Grunewald
Group Research and Transfer, DIN Deutsches Institut für Normung, Berlin, Germany
e-mail: sarah.koehler@din.de

C. Grunewald
e-mail: christian.grunewald@din.de

P. Rosa and S. Terzi (eds.), *Circular Approaches for the Ecodesign, Repair and Remanufacturing of Car and Mass Electronics*,
PoliMI SpringerBriefs, https://doi.org/10.1007/978-3-032-04301-6_11

1 Introduction into Standardization

What country is on your bucket list? Traveling the world is becoming more and more attractive, whether it is for vacation, business trips, to experience other cultures or to visit friends and family. Traveling as we do today was not always so easy. In particular, traveling within the European Union (EU) by train, plane or car is becoming easier and easier. Traveling by train within the European Union is only possible thanks to a series of technical agreements that enable the interoperability of trains and cross-border routes in different countries. A lot of things are taken for granted these days, but many people are not aware that there are rules which are applied in the background. For example, if you are involved in an accident in another European country, e.g. on vacation or at work, you certainly expect the same first aid as at home. Standards can help to achieve this. The European standard EN 1789 specifies requirements for the design, testing, performance, and equipment of road ambulances used for the transport, monitoring, treatment, and care of patients [1].

Another journey that is only possible as we know it today, is the door-to-door delivery process of products you ordered online. Especially products produced in eastern countries need to travel a long way but often don't take more than 5–7 days to reach their destination. Logistics and global trade are excellent examples to show how standardization can help. The international standard ISO 668 specifies classifications and dimensions for freight containers—known as ISO containers [2]. Around 250 million ISO containers travel around the world every year.

2 General

Standardization is a regulated process to find generally valid solutions for different problems or situations. As a result of this process, in which different stakeholders come together, discuss different approaches, and reach compromises, a standard document is published. The process of standardization is defined as follows: "Standardization[1] is the activity of establishing, with regard to actual or potential problems, provisions for common and repeated use, aimed at the achievement of the optimum degree of order in a given context" [3]. When thinking about global trading, cost saving, and promoting innovations, standardization is a great tool. Innovations and the marketability of new ideas can be supported by standardization.

Standardization is becoming increasingly important not only at national level, but also at European and international level by agreeing on terminologies, methods, requirements, characteristics, etc. in specific areas to make a product, process or service fit for its purpose. The process of standardization can be used to e.g. develop a generic language that is understood by all.

[1] Standardization covers all types of standardization documents and is used here in general manner.

3 Standardization Organizations

Standardization is a process that takes place not only at national level, but also in a European context and at global level. National standardization bodies (NSBs), which are members of the European and international organizations, delegate representatives to carry out the standardization work at European and international level and the management of these committees is divided between the member NSBs.

At national level, there are different structures and standardization bodies in different countries that carry out the practical implementation of standardization. Within the national standardization bodies different stakeholders representing different organizations from industry, commerce, the public sector or research, work together on the development of national standards, resulting in, e.g. DIN standards, BSI standards, UNI standards etc. All stakeholders can participate in this work, including manufacturers, consumers, businesses, research institutes, public authorities, and testing bodies.

Looking in the European context, standardization takes place to develop European standards, that are valid and accepted within the European Union. Around 85% of standardization projects today are in an international or European context [4]. There are three different main standardization organizations that coordinate and organize the standardization work at European level—*CEN* (Comité Européen de Normalisation, engl: European Committee for Standardization), *CENELEC* (Comité Européen de Normalisation Électrotechnique, engl: European Committee for Electrotechnical Standardization), and *ETSI* (European Telecommunications Standards Institute). The individual responsibilities depend on the focus of the standardization work. Standards developed within the EU, European standards—EN standards—must automatically be adopted by member states of the EU.

When thinking about activities like global trading, these things are easy to implement because of the process of international standardization. On international level, mainly the organizations *ISO* (International Organization for Standardization), *IEC* (International Electrotechnical Commission), and *ITU* (International Telecommunication Union) are responsible for international standardization work. In contrast to European standardization, there is no obligation to adopt international standards in national standards. Figure 1 summarizes the standardization structure at European and international level and displays the example of Germanys structure as part of it [4].

European and international standardization works according to the principle of delegation, which means that national standardization bodies send a selection of experts to the technical committees (TC) and working groups (WG) at CEN/CENELEC to develop European standards or ISO/IEC to develop international standards. In order to represent national positions at European and international level, so-called mirror-committees are set up and coordinated by the national standardization bodies. In these national mirror-committees, the work and existing results of corresponding European and international standardization committees are discussed, and a national position is developed [5].

Fig. 1 DIN as part of the European and international standardization structure. © DIN

Standards are given a five-digit number. At international level, the numbers have the suffix ISO, at European level the suffix EN. National standards have the addition of the national code, e.g. DIN XXXXX for German standards. If an international standard has been adopted at European level, the standard is referred to as ISO EN XXXXX with the same five digits. European standards must be adopted by all European member states, which is why the standard is designated at national level with the national suffix, e.g. ISO EN DIN XXXXX or EN DIN XXXXX.

As a result of the standardization process, a document is developed and agreed upon. Standardization documents can provide rules, guidelines or characteristics for activities or their results. There are different types of standardization documents.

4 Standard

A standard reflects the state of the art and should be based on the consolidated results of science, technology, and experience. The document specifies requirements for products, services and/or processes. A characteristic of a standard is that it is developed by consensus and approved by a recognized body or organization. All interested stakeholders should be present and involved to agree on e.g. requirements, common processes or language, etc. [4].

How are standards developed?

Before experts can discuss the latest developments on a specific topic and start to standardize, a proposal for a new standard is the beginning of the (national) standardization process. A proposal can be submitted by anyone to the responsible NSBs in each country. The responsible standardization committee on national level and its experts start the discussion for the need for a standard on that subject once the proposal has been received. In addition, they discuss the funding of the project and whether the work should be carried out at national, European or international level. Once the proposal has been accepted, the standardization work can start and the experts, who are active in the responsible standardization committee, discuss the content of the potential standard.

Figure 2 shows the standardization process with the example of DIN [4].

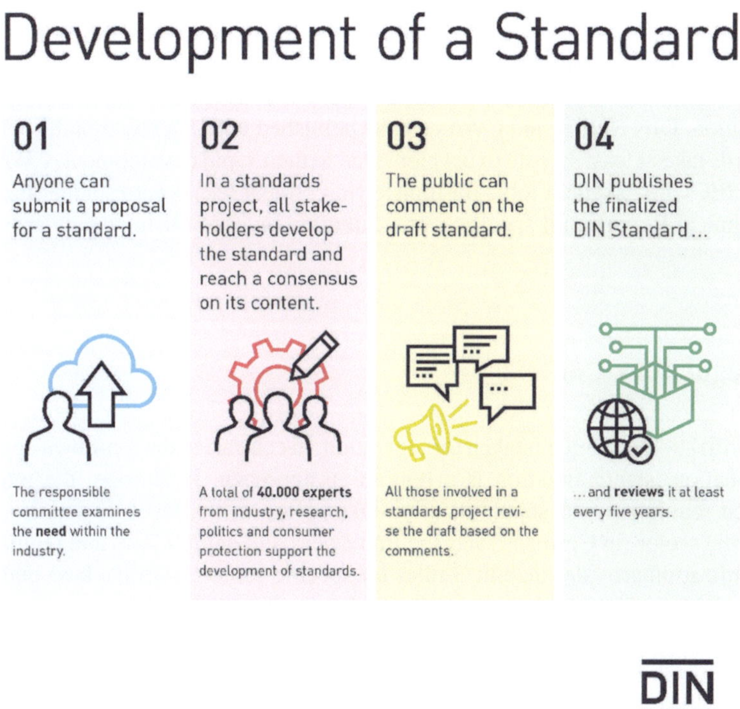

Fig. 2 The standardization process of DIN standards. © DIN

5 Other Standards Deliverables

Besides standards there are other documents which can result from the standardization process.

A *Technical Report* (TR) is a document that contains information on the status of standardization, including data, results etc. from standardization projects.

A *specification* is also a document which reflects the state of the art and specifies requirements for products, services and/or processes. In contrary, a specification does not require a full consensus, and not all stakeholders have to be involved. Specifications can be developed within the normal structure of a technical committee. These documents are called *Technical Specifications* (TS). However, specifications can also be developed outside the normal structure, e.g. within a workshop that aims to develop a standardization document such as a *DIN SPEC*, which is a specification at German level, a *CEN Workshop Agreement* (CWA) at European level or an *International Workshop Agreement* (IWA) at international level. These types of standardization documents can be used as strategic instruments for a quick and easy establishment and dissemination of innovative solutions to the market through standardization. DIN SPECs and CWAs can be published within a few months, whereas standards take at least 3 years to develop. Due to their rapid development, CWAs and DIN SPECs are excellent tools for transferring project results into standardization.

Technical Reports and specifications can often be used for later standardization work.

6 Regulatory Context

Standardization is often mistaken for regulation. In contrast to the top-down approach in regulation, standardization is a bottom-up approach. In Europe, the interplay between regulation and standardization takes place at the level of details. While the public sector (in Germany) sets the framework and defines laws and regulations, standardization provides technical rules for specific subjects. Unlike laws and regulations, the use of standards is generally voluntary. Standards can become mandatory as an exception if they are referred to as in laws or regulations. In this case, laws and regulations refer to "generally accepted rules of technology"—e.g. standards—for achieving the objectives. Standards are also often used in contracts and making them mandatory. Standardization can support and relieve the state regulatory authorities by specifying the legal obligations [4].

7 Standardization in Research Projects

It is crucial for a Research and Innovation project to know the state of the art in the areas relevant for or connected to the project. Research projects often deal with new ideas, new solutions or innovations. As standards represent the state of the art and reflect the market in specific areas, they could be used as a basis for the development of project results. Due to that it is essential to have an overview of the standardization landscape related to the project. Current technical rules can be applied for the development of new innovations, e.g. to ensure the interoperability with existing solutions. In addition, this knowledge helps research projects to address/ tailor their results to current market requirements.

Furthermore, by taking into account existing standardization documents and developing new solutions, awareness is raised on where standardization is still needed. Taking this a step further, the gaps in standardization can potentially be filled by project results, which is a great opportunity to disseminate results. Standardization can support the dissemination and exploitation of results and thus the implementation of innovations on the market through standardization activities, e.g. by contributing to ongoing standardization projects at national, European or international level, or by developing new standardization documents [6].

8 Standardization Within CIRC-UITS

For **CIRC-UITS** in particular, standardization plays an important role and has been taken into account since the beginning of the project. Different subtasks have been carried out and their results are described in the following subsections.

9 Standardization Research

An overview of the standardization landscape should raise awareness among the project partners on what already exists on the market and that project developments can be based on the state of the art. The standardization landscape provides the basis for further standardization activities of the project.

Keywords were collected within the consortium to enable a standards research tailored to the topics of the project. Based on the keywords given by the consortium, which were divided into four categories, the following existing standards and ongoing standardization activities were identified:

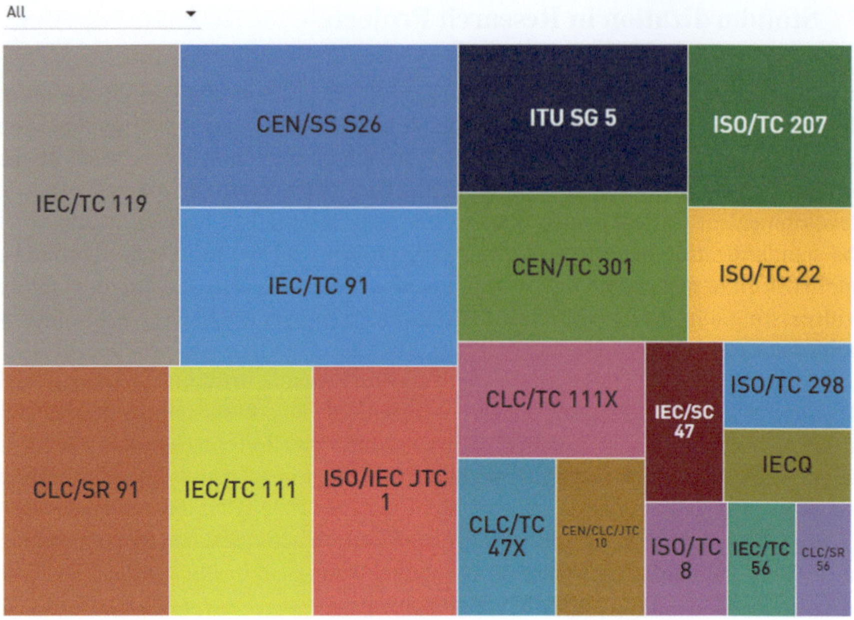

Fig. 3 Standardization committees at international and European level developed five or more standardization documents included in CIRC-UITS' standardization landscape. The size of the boxes represents the number of standards developed by the corresponding committees

The collected keywords were used to search the standards database Nautos[2] for existing standardization documents, standards under development, and technical committees (TCs) related to CIRC-UITS. By using the Nautos database, standards from European standardization organizations such as CEN, CENELEC, and ETSI as well as international standardization organizations, such as ISO, IEC, and ITU are included in the standardization landscape. Overall, around 400 standards were identified as project relevant. The CIRC-UITS standardization landscape shows the relevance of the project's subject both at international and European level. More than half of the standards relevant to the project are developed at international level. Figure 3 shows the most relevant international and European standardization committees [6].

The most relevant TCs are those that deal with topics such as printed electronics, environmental management, electronics assembly technology, information technology, etc. Table 1 lists the committees and their titles and gives more information about as relevant identified TCs.

Several standards developed by these TCs deal with different subjects of CIRC-UITS in different ways. Topics such as sustainability, circularity, and environmental management in relation to products and processes in the automotive and electronic mass sector are of particular interest.

[2] https://nautos.de/, DIN internal databases to perform the research of standards, standards under development, and further standardization activities.

Table 1 Standardization committees at international and European level and its title

Standardization committee	Title of standardization committee
ITU Study Group 5	*Environment and circular economy*
ITU Study Group 17	*Security*
IEC/TC 119	*Printed electronics*
IEC/TC 91 & CLC/SR 91	*Electronics assembly technology*
IEC/TC 111 & CLC/TC 111X	*Environmental standardization for electrical and electronic products and systems*
ISO/TC 207 & CEN/SS S26	*Environmental management*
ISO/TC 22 & CEN/TC 301	*Road vehicles*
ISO/IEC/JTC 1	*Information technology*
IEC/SC 47 & CLC/TC 47X	*Semiconductor devices & Semiconductors and Trusted Chips Implementation*
CEN/CLC/JTC 10	*Material efficiency aspects for products in scope of Ecodesign legislation*
ISO/TC 298	*Rare earth*
ISO/TC 108	*Mechanical vibration, shock and condition monitoring*
IEC/TC 56 & CLC/SR 56	*Dependability*
IECQ	*IEC quality assessment system*

Relevant standards at international level are also developed by the global association *IPC*, which supports OEMs in the context of electronics and by developing standards. (IPC) There are currently more than 15 standards published that relate to printed boards in terms of electronics, so IPC's impact on PCB standardization is enormous. IPC standards are not included in Fig. 3. Figure 3 only includes technical committees at European and international level that have developed five or more standardization documents, i.e. committees that have only recently been established and have therefore not yet published any or only a few standards are not included here.

There are two recently established standardization committees and their standardization work that are of great importance to CIRC-UITS. Regarding the topic of circular economy, the standardization committee *ISO/TC 323* 'Circular Economy' was established at international level in 2018. TC 323 published four ISO standards in May 2024, in which terms and definitions as well as assessments in the context of circular economy are specified [7].

- ISO 59004:2024—*Circular economy—Vocabulary, principles and guidance for implementation*
- ISO 59010:2024—*Circular economy—Guidance on the transition of business models and value networks*
- ISO 59020:2024—*Circular economy—Measuring and assessing circularity performance*
- ISO/TR 59032:2024—*Circular economy—Review of existing value networks*

- ISO 59040:2025—*Circular economy—Product circularity data sheet*

The document ISO/CD TR 59031—*Circular economy—Performance-based approach—Analysis of cases studies* is under development.

Five of these ISO standards were also considered within CIRC-UITS as soon as they were published or the draft available. The international work is mirrored at European level within the *CEN/TC 473* 'Circular Economy' [8], which is also of great importance in the context of achieving a circular economy in various industry sectors.

Regarding the industry sector, the topic of the "digital product passport" (DPP) is currently very interesting for the entire industry, not just for the automotive sector, but in a general context. Triggered by political actions like the Ecodesign for Sustainable Products Regulation (ESPR), the Circular Economy Action Plan (CEAP), and the Battery Regulation, the European technical committee *CEN/CLC/JTC 24*—DPP was established in 2023. The initiative aims to enhance the transparency, quality, and geostrategic resilience. The European Commission has proposed a draft standardization request to create a clear concept for DPPs by defining cross-sectoral product data models. Issues such unique identifiers, data carriers, links between physical and digital representations, access rights, interoperability, data processing, storage, authentication, and security should be addressed in future standardization documents. Standardization documents are currently under development [9].

The following are examples of standards on various topics that are relevant to the CIRC-UITS project.

The CIRC-UITS report D3.1 on "S&C assessment & advisory methodologies" [10] provides an overview of existing methodologies to evaluate sustainability and circular impacts. In this context the following standards have been mentioned and partially used within CIRC-UITS:

- ISO 15686–5:2017—*Buildings and constructed assets—Service life planning—Part 5: Life-cycle costing*
- ISO 14025:2006—*Environmental labels and declarations—Type III environmental declarations—Principles and procedures*
- ISO 14040:2006—*Environmental management—Life cycle assessment—Principles and framework*
- ISO 14044:2006—*Environmental management—Life cycle assessment—Requirements and guidelines*
- EN 15804—*Sustainability of construction works—Environmental product declarations—Core rules for the product category of construction products*
- CLC/TR 45550:2020—*Definitions related to material efficiency*
- EN 45554:2020—*General methods for the assessment of the ability to repair, reuse and upgrade energy-related products*

In the context of environmental management, the *ISO 1404X series* is of interest. The standard series provides principles, requirements, conditions etc. for the life cycle assessment of products, which is a key component of the CIRC-UITS project.

In the CIRC-UITS project, the Life Cycle Costing approach is used as described in *ISO 15686-5* for economic impact assessment.

The European standard series *EN 4555X* plays an important role in supporting the establishment of a circular economy and describes, among other things, methods for the assessment of the ability to repair, reuse or recycle energy-related products. Within CIRC-UITS the standard document *EN 45554* (EN 45554, 2020), which describes methods for the assessment of the ability to repair, reuse, and upgrade energy-related products, has been used and was applied.

Additional standards that are particularly relevant to the industry are listed below:

- VDA 232–101:2015—*Global Automotive Declarable Substance List (GADSL)*
- VDA 900–100:2022—*Guidance for Conducting Life Cycle Assessment Studies of Passenger Cars*
- EN ISO 9001:2015—*Quality management systems—Requirements*
- ISO 21750:2006—*Road vehicles—Safety enhancement in conjunction with tyre inflation pressure monitoring*
- ISO 23664:2021—*Traceability of rare earths in the supply chain from mine to separated products*
- IEC 62321-3-4:2023—*Determination of certain substances in electrotechnical products—Part 3–4: Screening—Phthalates in polymers of electrotechnical products by high performance liquid chromatography with ultraviolet detector (HPLC–UV), thin layer chromatography (TLC) and thermal desorption mass spectrometry (TD-MS)*

This is not a complete list of standards. More details about the standardization landscape—relevant standardization committees, standards that have been used within CIRC-UITS and a full list of standards—can be found in the CIRC-UITS Deliverable D5.2 "report on standardization activities" (1st version)[3] [6].

Partial overview of the European regulatory context on circular economy related to CIRC-UITS

In recent years, the European Union has adopted several regulatory relevant documents that should be taken into account in achieving a circular economy in the coming years. The next subchapter highlights some of the regulatory documents that should be considered when developing new and innovative solutions in the context of the CIRC-UITS project. However, this is only an excerpt from the existing regulatory documents and the list is not exhaustive.

In 2019, the European Commission adopted the *European Green Deal* [11] to create a framework for the transformative change towards the first climate neutral continent by 2050. To achieve these goals, various European directives and regulations have been developed and adopted in recent years.

In 2020, the *Circular Economy Action Plan* [12] was adopted by the European Commission. The plan focuses on the design and production for a circular economy to ensure that resources used remain in the EU economy for as long as possible.

[3] Status: June 2024.

The plan announces initiatives to support the transition to a circular economy by taking into account the whole life cycle of products. The sectors in which the most resources are used and therefore the potential for circularity is high are in the focus: electronics and ICT, batteries and vehicles, packaging, plastics, textiles, construction and buildings, food, water, and nutrients. To support the transition towards a circular economy, also on consumer level, the European Commission adopted the *Right-to-Repair Directive* [13] in 2024, which creates a framework for consumers regarding the right to easy repair goods. The directive aims to ensure cost-saving for consumers as well as waste reduction by supporting sustainable consumption and the repair of products instead of replacing them with new ones. The directive contributes to the ambitions set out in the European Green Deal. As part of the Circular Economy Action Plan, the *Ecodesign for Sustainable Products Regulation (ESPR)* [14] entered into force in 2024. The regulation aims to improve sustainable aspects, e.g. circularity or energy performance of products placed on the EU market. The ESPR describes a framework to set ecodesign requirements for different product groups.

In 2023, the European Commission proposed the *Critical Raw Material Act Regulation (CRM Act)* [15] as part of the European Green Industrial Plan, that builds on previous initiatives and complements the efforts of the European Green Deal. The CRM Act aims to the access to secure, diversified, affordable, and sustainable supply of critical raw materials in the EU.

To specify the technical implementation of European legislations, the European Commission can request standardization deliverables from the European standardization organizations. At the online database eNorm[4] standardization requests adopted by the Commission and addressed to the European standardization organizations are listed. As standardization documents reflect the state of the art and often specify the legal obligations, it can support the state regulatory authorities. In this case, upcoming regulatory plans can influence the standardization landscape.

More information on regulations can be found online by visiting the EU website.[5]

10 Standardization Needs Identified Within CIRC-UITS

Different standards have been used within the project. However, during the development of the project results, aspects were identified that are currently not addressed in the standardization landscape and the need for further standardization work was identified. To identify existing gaps in current standardization work in a systematic manner, a standardization potential workshop was held at the beginning of the second year of the project. Many innovative ideas were collected by the project partners, were sorted and filtered by the different categories given during the standards research

[4] https://ec.europa.eu/growth/tools-databases/enorm/standardisation_requests?language=en&triggerSearch=true.

[5] https://eur-lex.europa.eu/homepage.html?lang=en.

(see subclause 0). Afterwards the ideas were presented, ranked, and prioritized by the project consortium [6].

Following this procedure and monitoring the development of project results and the direction the project evolved, the following standardization ideas have been selected for further actions within CIRC-UITS:

Activities related to digital product passport

In the context of the *CEN/CLC/JTC 24—DPP*, the CIRC-UITS project identified a need regarding DPP in the context of printed circuit boards. Currently, it is very difficult to gather all information to disassemble PCBs the right way and support the reuse of components or the whole PCB. With a structured way of collecting and storing all relevant information and data of PCBs across the whole value chain, stakeholders like OEMs, recyclers, and manufacturers in the electronic industry can benefit by reaching a level of security and compliance, and the circularity approach can be supported. The CIRC-UITS project identified this need for standardization and has decided to take action in this area.

Activities related to the assessment to repair products

To support the transfer of circularity of products, an important approach is also to include the first phase of product development—the design-phase—into this approach. Even at this stage, design factors of products, e.g. PCBs or products in which PCBs are installed, can vary (e.g. the material, connections between components, size of products, etc.), which can be helpful to increase the possibility of circularity. To compare different designs and to assess the ability to repair, reuse or recycle a product, clear and standardized methods are essential. There are some standards which address methods to assess the ability to e.g. repair energy-related products (e.g., EN 45554), but the assessment methods are very complex, and the standards are difficult to apply for specific product categories. Since the project addressed this topic intensively, standardization activities in this direction have been initiated.

Contribution to ongoing standardization work related to rare earths

Rare earths are used in mass electronics and in the automotive sector, particularly in the semiconductor industry. Since the intensive rethinking of resource utilization, the reuse of rare earths has also become increasingly important. At international level, *ISO/TC 298* deals with rare earths in the context of mining, concentration, extraction, separation, and conversion, as it is essential for sustainable manufacturing and further production processes [16]. As rare earths can be used in semiconductors and CIRC-UITS is concerned with improving circularity in this sector, the project is interested in contributing to ongoing standardization work related to rare earths.

In addition, the following ideas were identified and selected as relevant, but could not be pursued in the context of standardization within the CIRC-UITS project due to limited resources:

Definition of additional Key Performance Indicators (KPIs) for sustainability

Since the standardization work of general aspects in the field of circular economy just started, with the first standards published in 2024, definitions and the calculation of KPIs in this field are still lacking. Within CIRC-UITS, social and circularity KPIs related to a repairability assessment of products were discussed and developed. These results can be used for future standardization work and can build the basis for discussion on this topic.

Repairability and a possibility to disassemble parts

Especially in the electronics sector disassembling a product is often very difficult. But when it comes to e.g. reuse, repair or recycling of a product, disassembling of a product is necessary and unavoidable. A DPP specifically for PCBs can support the evaluation of the ability to disassemble parts, but the implementation of disassembling components in a product needs to become possible. The CIRC-UITS project identified a gap in the standardization landscape related to specific methods for the repair and disassembly of products and parts of it. Methods need to be provided in a standardized way. A standardization activity is currently underway outside of the CIRC-UITS project to initiate the development of a German specification—VDE SPEC—entitled "Circularity check". The planned standardization work aims to close the gap between the Right-to-Repair, Energy Labels, Repairability Scores, Ecodesign for Sustainable Products Regulation, Data Act, and Digital Product Passport (Fixfirst). It will be evaluated during the project lifetime if this standard document could benefit from CIRC-UITS results.

11 Standardization Activities Within CIRC-UITS

Standardization can be a great tool to ensure the long-term dissemination and exploitation of project results and thus the implementation of innovations on the market. In general, there are different options to connect to standardization within research projects, depending on the maturity of the project results, the ongoing standardization work in technical committees at European and international level, and the relevance of specific topics. Within CIRC-UITS, different activities have been carried out and some are still ongoing, to come one step closer to closing the identified gaps in the standardization landscape, based on the identified standardization ideas.

Activities related to digital product passport

The CIRC-UITS project has applied to the European standardization committee *CEN/CLC/JTC 24*—DPP for a liaison in order to participate in the ongoing standardization work with regard to DPP. The application was approved by the JTC leading to the project becoming a liaison organization. The project sent one representative to the TC in general and one to the *working group 4—Interoperability framework*.

The representatives can attend the WG meetings and get access to the current standard drafts. Through the liaison the project has a great advantage in knowing which standardization documents are coming soon. They can also have a direct exchange with the standardization experts, to ensure that the results of CIRC-UITS could be transferred directly into standardization.

In addition to the direct connection to standardization work through the technical committee responsible for the DPP, the CIRC-UITS partners have decided to support the standardization work surrounding the DPP by initiating the development of a pre-standardization document and have therefore applied to hold a CEN/CLC Workshop. Currently, a *CEN/CLC Workshop Agreement* (CWA) with the title "Enabeling Circular Economy Practices: Repair and Recycling of PCBAs" is under development. The document will define requirements and recommendations for recycling and repair aspects for printed board assemblies (PBAs) and could provide the basis for the repair and recycling related section in a future digital product passport for PCBAs. The document excludes the definition of an IT infrastructure and will be orientated on the current developments of CEN/CLC-JTC 24—DPP. The workshop to develop the CWA was formed in January 2025. The workshop and the planned CWA will solve the issue of the lack of information in the electronics sector about materials, components, and whole PCBAs embedded in current products in any sector. Not part of the CWA is the analysis of existing data and/or their completion [17].

The CWA will probably be developed between January and July 2025. The publication of the CWA is planned by the end of the CIRC-UITS project in Dec 2025 and can be downloaded from the CEN/CLC *CWA download area* free of charge.

Activities related to the assessment to repair products

The project applied the EN 45554 in a specific use-case in the automotive sector. Some difficulties arose during the application, e.g. a comparison of the same product, but from different manufacturers, does not seem possible or at least very difficult. In order to provide the results of the repairability assessment based on the EN 45554 to the public and provide guidance for the application of the standard, the project decided to develop a *CEN Workshop Agreement* (CWA) with the planned title "Use-case for the application of the EN 45554 in the automotive industry". The document will describe the application of the EN 45554 from the perspective of manufacturers in the automotive industry. The document will provide useful guidelines to better understand the application of the EN 45554, and improvement ideas and recommendations for future standard revisions will be given. The CWA will be developed in close collaboration with the European standardization committee *CEN/CLC/JTC 10—Material efficiency aspects for products in scope of Ecodesign legislation*—and its *working group 3—Ability to repair, reuse and upgrade energy-related products*, that is responsible for the development and publication of the EN 45554. The CWA will probably be developed between January and July 2025. The publication of the CWA is planned by the end of the CIRC-UITS project in Dec 2025 and can be downloaded from the CEN/CLC *CWA download area* free of charge.

Possible contribution to ongoing standardization work related to rare earths

The international committee *ISO/TC 298—Rare earth* and its work program is relevant for CIRC-UITS. *Joint working group 6* deals with the accepted work item *ISO/AWI 24961—Rare earths and lithium sustainability across the value chain: concentration, extraction, separation, conversion, recycling and reuse* [18]. CIRC-UITS may contribute to the field of environmental sustainability for the recycling, reuse, and end-of-Life phases. In particular, the consortium has a lot of expertise to contribute to the context of environmental and circularity criteria and identifying new criteria related to the reuse of components. An Italian project partner is interested in participating in this standardization project. Contributing to ongoing standardization work can have a great impact and can support the transfer of project results onto the market.

12 Conclusion

CIRC-UITS is a great example of how standardization as part of research projects can be useful in the development of innovative solutions. By applying standardization documents in various areas that reflect the state of the art, it is possible to learn from them and benefit from existing technologies. Active participation in the standardization community facilitates the market transfer of methods and results developed in CIRC-UITS. It ensures that European values are more firmly anchored in standards.

In the context of supporting a circular economy, e.g. through the development of recycling, reuse, repair methods etc. or the development of completely new approaches, it will become increasingly important to become active and help shape future standardization documents.

List of Abbreviations

AWI	Accepted Work Item
BSI	British Standards Institute
CEAP	Circular Economy Action Plan
CEN	Comité Européen de Normalisation (engl: European Committee for Standardisation)
CLC	CENELEC—Comité Européen de Normalisation Électrotechnique (engl: European Committee for Electrotechnical Standardization)
CRM	Critical Raw Material
CWA	CEN Workshop Agreement
DIN	Deutsches Institut für Normung e.V. (engl: German Institute for Standardization)
DPP	Digital Product Passport

EN standard	European Standard
ESPR	Ecodesign for Sustainable Products Regulation
ETSI	European Telecommunications Standards Institute
IEC	International Electrotechnical Commission
IECQ	IEC Quality Assessment Center
ISO	International Organization for Standardization
ITU	International Telecommunication Union
IWA	International Workshop Agreement
JTC	Joint Technical Committee
JWG	Joint Working Group
NSB	National Standardization Body
SC	Subcommittee
SG	Study Group
SR	Reporting Secretariat
SS	Sub Sector
TC	Technical Committee
TR	Technical Repo
TS	Technical Specification
UNI	Ente italiano di normazione (engl.: italien standardization organization)
WG	Working Group

References

1. EN 1789+A1:2023. Medical vehicles and their equipment—Road ambulances
2. ISO 668:2020. Series 1 freight containers—Classification, dimensions and ratings
3. EN 45020:2006. Standardization and related activities. General vocabulary (ISO/IEC Guide 2:2004)
4. DIN (2024). https://www.din.de/de. Last access: 9 Dec 2024
5. Hartlieb B, Hövel A, Müller N (2016) Normung und Standardisierung—Grundlagen
6. Köhler and Grunewald (2024) CIRC-UITS research project, D5.2 "report on standardization activities" (1st version)
7. ISO/TC 323. https://www.iso.org/committee/7203984.html. Last access: 9 Dec 2024
8. CEN/TC 473. https://standards.cencenelec.eu/dyn/www/f?p=205:7:0::::FSP_ORG_ID:3326686&cs=1FB6C33369B0F6BEA8BA7EFC7B26D4309. Last access: 9 Dec 2024
9. CEN/CENELEC/JTC 24 (2024) Business plan CEN/CLC/JTC 24 digital product passport—framework and system. Available at: https://standards.cencenelec.eu/BPCEN/3342699.pdf
10. Fontana et al. (2024) CIRC-UITS research project, D3.1 Report on "S&C assessment & advisory methodologies" (1st version)
11. European Commission (2019) Factsheet: what is the European green deal. https://ec.europa.eu/commission/presscorner/api/files/attachment/859152/What_is_the_European_Green_Deal_en.pdf
12. European Commission (2020) A new circular economy action plan. https://eur-lex.europa.eu/legal-content/EN/TXT/?qid=1583933814386&uri=COM:2020:98:FIN
13. European Commission (2024a) Directive on common rules promoting the repair of goods. https://eur-lex.europa.eu/legal-content/EN/TXT/?uri=CELEX%3A32024L1799

14. European Commission (2024b) Regulation on establishing a framework for the setting of ecodesign requirements for sustainable products. https://eur-lex.europa.eu/legal-content/EN/TXT/?uri=CELEX%3A32024R1781&qid=1719580391746
15. European Commission (2024c) Regulation on establishing a framework for ensuring a secure and sustainable supply of critical raw materials and amending regulations. https://eur-lex.europa.eu/legal-content/EN/TXT/?uri=OJ:L_202401252
16. ISO/TC 298a. https://www.iso.org/committee/5902483.html. Last access: 9 Dec 2024
17. CEN-CENELEC (2024) https://www.cencenelec.eu/news-and-events/news/2024/workshop/2024-11-25_digitalproductpassport/. Last access: 9 Dec 2024
18. ISO/TC 298b. https://www.iso.org/standard/89047.html. Last access: 9 Dec 2024
19. DIN, DKE, VDI (2023) Standardization roadmap circular economy. https://www.din.de/resource/blob/906910/0d691bed63405ae85f281336ed71162c/standardization-roadmap-circular-economy-data.pdf
20. Fixfirst. https://www.fixfirst.io/circularity-check. Last access: 24 Oct 2024
21. IPC. https://www.ipc.org/. Last access: 9 Dec 2024

MIX
Papier aus verantwortungsvollen Quellen
Paper from responsible sources
FSC® C105338

If you have any concerns about our products,
you can contact us on
ProductSafety@springernature.com

In case Publisher is established outside the EU,
the EU authorized representative is:
Springer Nature Customer Service Center GmbH
Europaplatz 3, 69115 Heidelberg, Germany

Printed by Libri Plureos GmbH
in Hamburg, Germany